ENERGY

ENERGY
Ending the Never-Ending Crisis

Paul Ballonoff

CATO
INSTITUTE
Washington, D.C.

Library of Congress Cataloging-in-Publication Data

Ballonoff, Paul A.
    Energy : ending the never-ending crisis / Paul Ballonoff.
      p. cm.
    Includes bibliographical references and index.
    ISBN 1-882577-45-0 (cloth). — ISBN 1-882577-46-9 (paper)
    1. Energy policy—United States. 2. Public utilities—Law and legisla-
tion—United States. 3. Energy industries—Government policy—United
States. 4. Energy industries—Deregulation—United States. 5. Trade
regulation—United States. I. Title.
HD9502.U52835 1997
333.79'0973—dc21   97-42891
                                                    CIP

Printed in the United States of America.

CATO INSTITUTE
1000 Massachusetts Ave., N.W.
Washington, D.C. 20001

Contents

Preface

The Path of This Book

When I was young, my favorite poem was Robert Frost's "The Road Not Taken." Never afraid of being alone on the highway, I certainly followed several less traveled paths in the work that led to the text before you. The final destination, however, was unexpected. The book began as a study of the persistent failure of energy price forecasting and its attendant consequences. But it became difficult to understand a century of intellectual failure in an era when so much political, economic, and cultural analysis claims to have some "scientific" mooring.

Real science is tested by contrasting predictions with reality. When scientific predictions persistently fail to reflect what is known to be real, theory is reworked to reflect reality more accurately. That this has not occurred in the realm of energy policy and economics is astonishing. It requires explanation. It requires abandoning the well traveled roads, since they seemed to lead nowhere interesting.

Those relatively untraveled roads led me from energy forecasting to constitutional theory. First, it was necessary to understand the technical issues at hand. Having done so, it then became my duty to understand why and how an important area of public policy went so badly wrong. This led to an exploration of the political and legal context in which the policy arose.

To understand the traditional perspective, it was necessary to review the classic cases that laid the foundation for modern utility and energy law. My review led me to conclude that two constitutions exist: one written by the Founding Fathers and another arising from both historical accident and deliberate policy, not during the New Deal but a half century earlier, in the chaos that followed the Civil War and the subsequent rapid expansion westward.

What I discovered is that today's energy policy rests largely on a power that the federal government was never given; indeed, was *intentionally* not given. This book calls that power the "general patent power." The name is used to describe governmental allocation of special economic rights, a power that the federal government simply does not possess. The term distinguishes the general patent power from the more specific patent power—related to new inventions—which the federal government does indeed possess.

Acknowledgments

I want to acknowledge five individuals or institutions whose influence inspired me to write this book. First, as usual, but with more than the usual influence, must be my undergraduate instructors in microeconomics at the University of California at Los Angeles, principally Jack Hirshleifer and Harold Demsetz. Neither would have predicted my interest in the subject of this book. Even less in 1960 to 1964 could I have predicted the influence on me, in 1988, of Professor Demsetz's 1968 paper "Why Regulate Utilities?" (*Journal of Law and Economics,* vol. 11, April 1968, pp. 55–65).

The problem posed by this book is the same one posed by Demsetz, though the solution offered here is different. In some ways, Demsetz's solution is more radical than mine: he would seek a market alternative to utility regulation by substituting periodic auction of local utility rights, whereas I propose a much more ordinary market with no more than the same antitrust regulation applied in the rest of the economy. Demsetz implicitly accepts the existence of "natural monopoly" scale economies in utility costs. I believe the relative scale economies are not very large and in particular believe they do not amount to natural monopoly and, hence, I recommend different solutions to the problem of inefficient regulation. But I would not have reached my solution at all without having studied, in the 1980s, Demsetz's earlier analysis. I did so at the suggestion of Professor Hirshleifer, who always has been forthcoming with intellectual support for a former student short of knowledge but long on problems.

Two further acknowledgments arise from my decade of employment in state regulation, divided equally between the Kansas State Corporation Commission and the Illinois Commerce Commission. The 1980s were an intellectually exciting period to be working in the regulated energy industry. Kansas provided the practical experience, and Illinois the intellectual stimulation, to explore the problems discussed in this book. In Kansas I had the opportunity to read the literature on the theory of utility regulation, and encountered the works of Richard Posner. In "Natural Monopoly and Its Regulation"

(*Stanford Law Review*, February 1969, vol. 21, no. 3, pp. 548–643), Posner argues that utility regulation cannot improve on uncontrolled natural monopoly, assuming of course that such things even exist. Posner summarizes (page 549):

> This study has convinced me that in fact public utility regulation is probably not a useful exertion of government powers; that its benefits cannot be shown to outweigh its costs; and that even in markets where efficiency dictates monopoly we might do better to allow natural economic forces to determine business conduct and performance subject only to the constraints of antitrust policy.

In many ways this conclusion also summarizes the substance of the present book, at least regarding claims about the existence of natural monopoly as a basis for utility regulation. Unlike Posner, however, I provide the reader with data about utility costs and the possibility of multiple distribution grids to demonstrate that utility competition is indeed possible. In addition, I believe in the viability of the "antitrust option" for completely different reasons than those of Posner, who, though writing in a law journal, hardly discusses the legal foundations of regulation at all. Indeed, the analysis of the constitutional basis (or lack thereof) for regulation in this book is, so far as I can determine, an original contribution to the literature.

Another source of ideas in this book was the opportunity for development of concepts provided by the biennial European Meetings on Cybernetics and Systems Research at the University of Vienna, Austria. I first presented many of my ideas in papers prepared for the 1988 and 1990 sessions.

The last source was National Economic Research Associates, Inc., a firm for which I was a consultant. I first discussed several of the key ideas found here—especially those of Chapters 1 and 2—in projects for the firm's clients. I especially thank Michael Rosenzweig, a vice president of the firm in their Washington, D.C., office, for encouraging my work on some of these ideas.

The book would not have been written without the support of the Cato Institute. The interest of David Boaz and Jerry Taylor, responsible at Cato for publications and natural resource policy, respectively, made the book possible. Jerry Taylor's comments and suggestions, and Peter VanDoren's editing on behalf of Cato, made this a much better work.

Units of Measure

This book discusses a great deal of data. The energy industry also has its own terminology. I keep these as painless as possible by greater use of narratives than charts and tables, avoiding jargon, or explaining unique terms and processes of the industry when needed.

It is useful, however, to state up front a few conventions for the measures used. In general, I cite energy quantities in "quads," or quadrillions of British thermal units (Btu's). For comparison, a quad of energy is approximately the amount of energy of all types (electricity, natural gas, coal, fuel oil, etc.) consumed by residential utility users in New York and New Jersey in 1990. A quad is also a bit less than the consumption of all types of energy used to generate electricity, or a bit more than the total energy of all types used for transportation in the New England area in 1990.[1]

Source data stated in other units are usually converted here to quads. Many natural gas industry citations report natural gas reserves in trillion cubic feet (Tcf). At standard measures and conditions, natural gas is methane whose heat content is 1,000 Btu per cubic foot. Thus, 1 quad = 1 Tcf. Another commonly used measure in the natural gas industry is 1 billion cubic feet (Bcf). A Bcf is simply 1/1,000 of a Tcf, that is 1/1,000 of a quad, or 100 Bcf = 1/10 quad. Since conversion of values in Bcf's would result in rather large quantities of energy being stated in numbers that look like small fractions of a quad, I leave most data originally stated as Bcf in the original form. Some data were stated in barrels of oil equivalent (or BOE) contents of energy. I treat one BOE as containing 5.8 million Btu's per barrel.

All energy prices are stated when possible in dollars per million Btu (MMBtu), or for electricity, in dollars per kilowatt hour (kWh). Prices are in current dollars for the year represented by the price. This avoids any confusion that could be introduced by choice of base year or inflation discount rate when current year nominal prices are converted to "real" prices.

The views presented in the present book are my own. No one else is responsible for how I present them, nor for any errors of fact or omission.

---

[1]See Table 93, p. 164, and Table 94, p. 166 of U.S. Department of Energy, Energy Information Administration, *Supplement to the Annual Energy Outlook 1993.*

1. The Never-Ending Crisis

In 1871, the state of Illinois decided it needed to regulate commercial grain storage passing through the great Chicago warehouses. Shortly thereafter, Ira Y. Munn and his partner in a major elevator, George L. Scott, suffered the indignity of conviction for failure to comply with those new regulations.

Given the other events in American political history after the Civil War, Munn's criminal conviction may seem trivial, but the consequences—a constitutional stamp of approval on price controls, restrictions on market entry, and government micromanagement of business practices—continue to affect how government views and regulates business, especially those businesses we call utilities.

After the *Munn* case, near-absolute state control of various businesses was achieved through the establishment of state commissions. Those commissions were charged with controlling "natural monopolies," which, it was thought, would produce too little product at too high a price without government intervention. Massachusetts created the first state electric regulatory commission in 1887, and by 1934 all but eight states had such bodies. Natural gas services, which had existed in many cities from as early as 1816, also were brought under the control of these same bodies.

The Cult of the Bursting Bubble

The development of the U.S. oil and gas industry also has been accompanied since its inception by demands for government regulation. The rationale has two components. The first is the belief that a severe reduction in petroleum supply is always imminent. The second is the claim that petroleum markets also are monopolistic with the result that prices to consumers are too high.

Concerns about imminent shortages have existed almost from the very beginning of the petroleum industry in America. For example, the U.S. Revenue Commission said in 1866—only seven years after

1

the first oil well was drilled in Pennsylvania—that the country soon would need to rely on synthetic oil. In 1885 the U.S. Geological Survey said there was no chance of finding oil in California; in 1891 the Survey said there was no chance of finding oil in Texas or Kansas. In 1914 the U.S. Bureau of Mines said that the total future domestic production would be only 5.7 billion barrels. Today, we still have little need for synthetics, California has become a major producer, Texas and Kansas have remained major oil and natural gas producers on a world scale for nearly a century, and the total production in the country is more than 34 billion barrels since the 1914 forecast.[1]

Claims about imminent shortages have dominated debates and controlled national petroleum policy for the last century. They still appear almost daily about fuels that brighten our lights, power our vehicles, or heat our homes and industries. Yet over the same century energy supplies have increased faster than energy use, both world-wide and nationally. This has occurred despite major economic, political, and military disruptions often intended to produce the opposite result.

The idea that the petroleum industry is not competitive has a historical basis or, better stated, a political one recorded in American history. The first U.S. antitrust law—the Sherman Act—was passed in 1890. Its aim was to provide special restrictions over the commercial activities of the so-called "trusts," which had become so large by the late 19th century that public action was deemed necessary.

In 1906 Standard Oil Company was made the subject of a federal antitrust suit for conspiracy in restraint of trade. The trial court found for the government, and the Supreme Court upheld the result in 1911. The result was the dissolution of Standard Oil into numerous competing companies. Daniel Yergin, president of Cambridge Energy Research Associates and author of The Prize, sees this as a triumph of politics: "Public opinion and the American political system had forced competition back into the transportation, refining and marketing of oil." Indeed, even among the "experts," the success of the 1906 suit against Standard Oil colored subsequent perceptions of the industry.

---

[1]Cited from original federal agency sources at p. 142 in James T. Bennett and Thomas J. Dilorenzo, *Official Lies: How Washington Misleads Us* (Alexandria, Va.: Groom Books, 1992).

2

Thus, in the 1970s, when federal oil price regulation was limiting supply and foreign companies were taking advantage of the situation to raise prices, domestic politics was still obsessed with the oil industry's "obscene profits." That rhetoric drove national policy, despite the fact that the industry was even less concentrated in the 1970s than after the dissolution in 1911, and that industrial profit margins were in fact nearing historic lows.[2] The same pattern repeated itself in the winter of 1996 when environmental regulations were responsible for a modest temporary spike in gasoline prices, bringing them to a level only a bit higher than normal for the season. Nevertheless, the price movement, occurring as it did amidst the heat of the 1996 political campaign, provided ample opportunity for political opportunists to decry corporate oil conspiracies.

Regulation Reconsidered

Thus, for more than a century, many public officials and academic specialists have had two views about energy markets, both of which were claimed to warrant government regulation: monopoly and imminent depletion of fossil fuels. The dominance of those views encouraged legislators to establish an elaborate system of energy regulation. The policies themselves, however, have consistently proved much worse for the country than any of the alleged conditions they were intended to ameliorate.

The purpose of this book is to challenge the intellectual premises and rationales for energy market regulation. Chapter 2 begins with one of the most persistent and widely believed myths of energy: that petroleum, natural gas, and uranium energy supplies are an inherently limited resource, a so-called finite resource, so precious that as each drop is used, cost and price necessarily skyrocket. The chapter demonstrates that the opposite is true. Energy production, including petroleum energy production, is a modern high technology product. Its economics are the same as those of other high technology industries; as technological knowledge increases, costs drop and supply increases. Many of the popularly cited measurement tools and

[2]Daniel Yergin, *The Prize* (New York: Touchstone Books, 1992), p. 110. See pp. 108–10 for discussion of the Standard Oil suit, pp. 111–13 for discussion of the changes in technology in the immediately following years, and especially pp. 656–59 for discussion of the allegations of "obscene profits" rhetoric of the Jacksonian Democrats, leading eventually to the energy laws of the Carter era.

3

statistics used to spread panic about petroleum supply are, in fact, indexes of an increasingly healthy market with falling costs. For example, the commonly cited falling reserve lives and falling reserve replacement rates are simply manifestations of a now common industrial process: just-in-time inventory management. As information technology increases ability to predict drilling successfully, drilling costs drop and the necessity for large current reserves falls. Therefore, inventories can be reduced (falling reserves, and also, falling "reserve lives"), leading to both lower costs and lower prices.

Chapter 2 ends with an examination of the cost structure of petroleum production in light of the requirements of a competitive market. In short, production is found to be highly competitive and the energy production industry has little tendency toward concentration. There is an increasing success rate in finding new reserves and developing new wells, leading to steadily decreasing costs. The structure of the industry is very much like the classic textbook example of a dynamic competitive market, using several factors of production to manufacture a technological product.

Chapter 3 examines similar issues in the context of electricity and natural gas utilities. Transmission and distribution of electricity and natural gas are not "natural" monopolies. A review of the structure of electricity and natural gas costs does not support the existence of large economies of scale. In addition, the history of the electric industry suggests that utilities themselves demanded regulation because of the existence of *too much* competition rather than too little. The belief that the distribution grid of wires (or pipes for natural gas) is a monopoly and an impediment to competition forgets the existence of six service "grids" (phone, electricity, gas, water, sewer, and cable) linked into most American households, each of which could compete with the others.

The final section of Chapter 3 asks whether other market failures concerning consumers' inability to compare present and future choices as well as the existence of environmental pollution necessitate the existence of regulation. The alleged difficulties of consumers do not stand up to scientific scrutiny while the magnitude of environmental externalities, though greater than zero, does not require the level of external taxes being proposed by energy planners.

Chapter 4 examines three major effects of energy market regulation that receive limited attention in the literature. First, utility regulation does not eliminate the existence of rents or excess profits; it simply transfers them to other firms or to the government itself in the form of taxes on utilities. Second, energy regulation has encouraged inefficient technologies (such as nuclear power and electric cars) and discouraged those technologies that would enhance efficiency (such as natural-gas-fueled vehicles). Finally, energy regulation inhibits innovation in the organization of natural-gas and electric utilities that would lower the costs of both for consumers. Traditional concepts of how engineering realities define industry segments no longer apply. The competitive advantage in electricity production may lie with a natural gas company. The competitive advantage in long-distance transmission of natural gas could lie with an electricity-generating company. Regulation prevents either from occurring.

Chapter 5 is a review of the dubious constitutional foundation of energy regulation. By the end of the 19th century, states were able to impose both price and entry regulation on services "affected with a public interest." In addition, the federal government itself regulated markets in territories. Despite this more than 100-year-old history, current state and federal regulation is probably unconstitutional. Chapter 5 argues that the courts should prohibit most current public utility regulation because the federal government is not empowered to enact such regulation and the states may do so only if such intervention does not restrict competition, which current price-and-entry regulation clearly does.

Markets should be left alone. If government is nonetheless compelled by the public to act, the least harmful intervention is a sensible and restrained antitrust regime no different from that which applies to every other arena of economic life. Therefore, Chapter 6 argues that the first-best utility regulatory policy is to eliminate utility regulation, period. But utility regulation was created for political purposes, and it may be impossible to remove politically. If that is the case (and the "common wisdom" that utility regulation cannot be eliminated may be as wrong as it is about other forecasts related to energy), then we should do the next best thing and turn utility regulators into antitrust courts of first resort. All authority for commissions to set prices and prohibit entry should be removed. Instead, regulators should be required to do no more than act as special antitrust courts for energy matters.

In practice, this means that if a competitive violation is found, utility regulators would encourage competitive entry and not create protected franchised markets. Rather than set price, regulators would act only when undue price discrimination is found. In such cases, regulators should ensure that price variations are not wildly different from those competitive markets would create. After all, excessive exuberance in the campaign to eliminate price differences would dampen the economic signals necessary to prompt investors and entrepreneurs to provide additional services. In the absence of a provable violation of the mechanisms of competition, utility regulators would not act, or need to act, at all.

Chapter 7 concludes with some final thoughts about the issues addressed herein. Why is it that what seems so self-evident to this analyst could appear so radical to others? Why does the myth of vanishing energy and natural monopoly continue to hold such a firm grip over otherwise intelligent and thoughtful people? While there are no easy answers to those questions, a combination of professional self-interest, the disease of collective thinking, and the blinding effect of ideology explains a lot.

This book should not be taken as advocating aggressive antitrust action as a substitute for the more complete controls applied by traditional utility regulation. Rather, antitrust should be viewed as the outer limit of constitutionally permissible regulation. An enumerated federal power, "the commerce clause" of the Constitution, whose purpose is to prevent the states from using their more general plenary power to interfere with commerce among the several states, is the basis for the use of federal power to prevent the states from creating or perpetuating local monopolies that interfere with interstate commerce.

The political lesson of this century is that the best way to reinvent government is to reduce what it does. But if electoral pressure forces government to act, we must constrain the intervention by establishing institutions that do the least harm.

2. Do Petroleum, Natural Gas, and Energy Markets Need to Be Regulated?

Despite the general support for free markets in the United States, Americans seem to believe that energy markets should be regulated. There are two main reasons for this. First is the belief that energy supplies are finite and that a depletion crisis is always imminent. Second is the belief that the petroleum, natural gas, and electricity industries are not competitive and that consumers accordingly pay prices that are often too high. This chapter discusses both issues, but particularly those concerning the petroleum and natural gas industries.

Petroleum and Gas: Technological Products, Not Finite Resources

The traditional view of energy supply is that energy is a finite resource that exists in a fixed supply. Obviously, if a commodity is in finite supply and then some of it is used, the total amount remaining must be less than the initial amount. With even more use, the total supply shrinks still more, until, eventually, an extreme supply shortage relative to demand exists. In essence, this traditional energy perspective views energy like buried treasure. We stumbled across its existence as a resource a little more than a century ago, went about finding it with the aid of various "treasure map" experts, and now believe that most of the treasure has already been dug up.

The treasure map theory of energy, enhanced with a bit of politics, becomes the following public policy prescription: "Energy is a vital finite resource, being consumed incessantly because of greed and market imperfections, stripping vitality from helpless, desperate neighbors and offspring. Soon, only rich people will be able to afford to use light bulbs. Accordingly, we must regulate energy! That way, the government can limit its use to prolong the supply and produce a fair distribution to everyone, before it all runs out."

7

The finite resource argument as applied to oil and gas has been made repeatedly for decades. The finite resource model is, of course, trivially true in the short run: every resource or production process is a finite resource in some sufficiently short time period. We can produce only as many computer chips today, for example, as current production capacity will allow.

In the longer run, however, we can build more chip factories. Therefore, while the total productive capacity for computer chips is fixed in the very short term to existing, known, "proven" capacity, over a longer time we can build new factories and expand the existing ones. We can invent new kinds of chips that require entirely different kinds of factories. Even over an intermediate time, we can increase the output of an existing factory by improvements in its design and operation. Thus, the production capacity of an industry is not an inherent, permanent limit on total capacity.

The same insights are applicable to the energy industry and explain why the finite resource model has failed to predict long-run supply or price for petroleum energy. Petroleum energy is a technological product. It differs little from computer chips or any other manufactured product in its basic economics. True enough, we cannot produce any more petroleum today than is available from existing known, connected, operating, proven wells. But we can drill more wells in the future, and we can drill them at lower costs as our knowledge increases. In addition, we can expand the known life of existing wells by adapting their operation as knowledge increases. And, if or when the cost of petroleum rises because of increasing scarcity, we can invent more efficient means of using that petroleum. Ultimately, we can even invent new forms of energy to substitute for petroleum, just as petroleum was itself a substitute for whale oil. Happily, all of this would happen naturally within the free market without any government direction.

As with any other product whose existence depends on technology, the cost of petroleum production depends on the state of technological knowledge. As technology progresses, the cost of energy production falls.[1]

---

[1] A more radical proposition argues that petroleum located deep in the earth's crust (30,000 feet) refills reservoirs located at much shallower depths (6,000 feet) and, thus, petroleum is an even less finite resource. See Malcolm W. Browne, "Geochemist Says Oil Fields May Be Refilled Naturally," *The New York Times*, September 26, 1995, p. B5.

The Computerization of Petroleum Manufacture

Technological change has been important in the petroleum and natural gas industries since their inception. When petroleum was first developed, oil was shipped in wooden barrels. Later, oil was shipped by rail tanker cars and underground pipelines. Both those innovations lowered the costs of oil manufacture. Initially, methane, the "natural gas" often found in oil wells, was a waste product. Then someone thought of using "useless natural gas" to illuminate streets. In the early 20th century, methane was manufactured in chemical-processing plants and distributed at low pressure in cast-iron pipe systems. Local distribution (of *manufactured* "natural" gas) became part of modern urban life. Discoveries of new oil and gas fields, coupled with improved pipeline and compressor design, made long-distance transmission of natural gas or liquid gasoline economic after World War II. The innovations lowered the price of natural gas but put local methane manufacturers out of business, or at least put their plants, mostly owned by local distribution utilities, into mothballs.

Computers have transformed petroleum exploration and development from a speculative enterprise into a space-age undertaking of managed knowledge. Historically, exploration and development of new fields were considered "risky" undertakings because of uncertainty about whether oil would actually be found or, if oil was found, how large the pool might be. When the analytical abilities of improved computers were adapted to oil exploration and development, the success rates for finding and drilling oil rose dramatically.

Today, instead of the pipe valve, the industry symbol is more appropriately a virtual-reality, full-color simulation of deep rock formations presented next to photographs of grinning "roughnecks" wearing cotton work shirts and bearing M.A.s in mathematical statistics from the University of Houston. Productivity of petroleum exploration and well development now depends on the efficiency of computers. Each advance in computer processing capability causes the supply curves for petroleum exploration, development, and production to fall by a noticeable percentage.[2] The cost of creating "natural" gas or petroleum drops, and the real market price

---

[2]See John Markoff, "Denser, Faster, Cheaper: The Microchip in the 21st century," *The New York Times*, December 29, 1991, p. F5, for support for the belief that innovations will continue to decrease the costs of computer computations.

moves accordingly. Major cost efficiency gains have been reaped by increasing computerization, most notably in the management of geological information.[3] Use of three-dimensional imagery of underground formations has improved the ability to locate precisely petroleum-bearing formations. More subtly, increased ability to amass knowledge about known formations has converted established producing fields, previously considered "old" finds, into major present sources of low-cost "new" supply.[4] Use of satellite sensing coupled with computer-enhanced imagery has greatly assisted the identification of promising geological formations.

The Information and Energy Revolution: Case Studies

An example of the profound effect that advances in information technology are having on the petroleum industry can be seen by considering the recent changes at Bow Valley Energy, a major oil and gas producing company headquartered in Alberta, Canada.[5] Bow Valley Energy was rife with interdepartmental conflicts about the ownership of particular bits of information used by each department. After the installation of a new data management system using personal computers for access from any department, those disputes evaporated.

Using the new data system enabled staff members to reduce significantly the processing times for many diverse functions of the company. Preparation of a document that previously took several

[3]See especially the many articles in the *Oil and Gas Journal*, for discussion of these issues. A few specific citations include the following: Stephen W. Knecht et al., "Case Study: How a Small Company Adopted 3-D Seismic Technology" (October 19, 1992), pp. 54–57; David C. Koger, "Landsat 6 Images to Provide More Remove Sensing Muscle" (February 1, 1993), pp. 51–55; D. B. Neff et al., "Technology Enhances Integrated Team's Use of Physical Resources" (March 31, 1993), pp. 29–35; A. D. Koen, "Tough Economics Remold Production R&D Campaign" (July 5, 1993), pp. 14–18; Davis W. Ratcliff, "New Technologies Improve Seismic Images of Salt Bodies" (September 27, 1993), pp. 41–49; Peter K. Trabant, "Seismic Stratigraphy, A Solution to Deepwater Drilling Problem" (September 27, 1993), pp. 50–56; Fred J. Pittard et al., "Horizontal Gas Storage Wells Can Increase Deliverability" (October 11, 1993), pp. 75–79; and Jerry W. Box, "Reorganization at Oryx Energy Focuses on Teamwork, Technology" (November 8, 1993), pp. 54–61.

[4]See Argonne National Laboratory, "An Assessment of the Natural Gas Resource Base of the United States," May 1988.

[5]"Bow Valley Uncaps a Well of Knowledge," *The Globe and Mail*, Toronto, Canada, January 18, 1994, p. B26.

months was done in weeks. Geologists who previously required weeks to prepare maps now let computers do the job, a change that allowed the geologists to spend more of their time thinking about geology. Others who needed the maps could now simply call them up on computer screens. Employment in the Calgary office of the company was reduced accordingly from 400 to 300 people, but total productivity increased. Additional cost reductions will occur because the company is expanding use of the same data system to its operations in London and in Indonesia.

Another interesting example comes from a study of the employment history and prospects of petroleum engineers in the United States:

> With modern computer technology, for example, PEs [petroleum engineers] can obtain and analyze massive amounts of data and present the results in clear tabular form abetted with color graphics. Reports typically can be produced in less time than formerly was needed just to punch cards to enter data into mainframe computers.[6]

The report showed that membership in the Society for Petroleum Engineers dropped by almost half from 1980 to 1990. In the two years from 1992 to 1994, new hiring by large operating companies dropped by nearly 20 percent, while employment by small operating companies increased by 10 percent, and by independent service companies by nearly 20 percent. The results are all compatible with the view that innovation drives cost by massively increasing productivity, which, in turn, drives employment trends in the industry. More specialized or innovative firms increase employment, while those firms whose business is in part the maintenance of inventory (the larger operating companies) decrease employment.

Consider also developments at one of the world's oldest producing fields, the Hugoton Field, located largely in western Kansas. One of the largest operating producers in that field is Occidental Petroleum Corporation, inheritor through corporate mergers of some of the oldest wells in that field. According to the corporation's 1992 *Annual Report*, "A project to automate OXY USA's more than 1,100 Hugoton gas wells, which is 70 percent complete, reduces operating costs by

---

[6]A. D. Koen, "Economics, Technology Rule Petroleum Engineers' Future," *Oil and Gas Journal*, March 28, 1994, pp. 23–27.

providing accurate and timely production data with a minimum of personnel."[7] The company went on to note that it has "lowered per-barrel production costs by improving operating efficiency at all levels, adding low-cost production to the reserve base, and divesting nonstrategic and marginal properties." In a rare show of preening for an otherwise terse business report, the subject is mentioned again on the same page in the caption to a fine, full-color photo of a producing well, presumably also located, judging from the visible blue sky in the photo, in the Kansas Hugoton.

Materials Technology and Improving Efficiency

Similarly, improved materials technology has led to vastly improved drill bits, which, in turn, reduce drilling costs by increasing drilling speeds, reducing breakage of bits, the number of types of bits required for different strata in a given well, and total labor and capital costs required to develop a particular well.[8] Innovations in the design, construction, operation, and movement of off-shore platforms, an increasingly important means of oil and gas recovery, are also reducing the total costs of petroleum production.[9] The diversity of technological advances that affect this is truly amazing: increased automation that both improves safety and reduces total weight on the platform, high technology materials like Kevlar used to protect surfaces, drill bits with cameras in them to show operators a real-time image of the drilling in action, horizontal as well as vertical drilling, corrosion resistant materials including composites used in platform construction, advances in computer technology to assess geological data, and so forth.

Wellhead Productivity: The Untold Story

Those increased efficiencies of oil extraction are reflected in the data available on petroleum exploration and development. According to the U.S. Department of Energy's Energy Information Administration (EIA), the number of active rotary drilling rigs in the United

[7]Occidental Petroleum Corporation, *Annual Report*, 1992, p. 9.

[8]See Peter Britton, "Offshore Oil, How Deep Can They Go?" *Popular Science*, January 1992 for a description of many new technologies including fiber composite materials for construction, better drill bits, horizontal drilling techniques, and the information management methods emphasized.

[9]A contributing factor to the decline of energy prices in the 1980s is the rapidly declining costs of computing power stemming from the introduction of personal computers.

States fell from a peak of 3,105 in 1982 to 1,010 by 1990, and to 775 in 1994.[10] Yet the average productivity (in millions of barrels of oil equivalent finds) of on-shore wells rose from 3.261 million barrels in 1970 to 5.338 million barrels in 1990, with similar productivity ratios holding into 1994.[11] The productivity of off-shore wells actually decreased a bit, from 13.0 million barrels of oil equivalent in 1970 to 11.4 million barrels of oil equivalent (MBOE) in 1990 and similarly through 1994. Between 1970 and 1990, however, productivity reached a low of 1.734 MBOE for on-shore wells in 1980 and 6.3 MBOE for off-shore wells in 1985.

Those measures suggest that rig productivity is inversely related to regulation. It was lowest when price controls were in effect and rose sharply immediately after even partial deregulation in 1985.[12] The success rate for exploration wells went from under 20 percent in 1970 to about 30 percent in 1990, with an industry average for all wells of 37 percent in 1994. The success rate for development wells rose from about 75 percent in 1970 to over 82 percent by 1994.[13]

Especially revealing is the measurement of barrels of oil equivalent found per successful foot drilled. This is a more useful measure than a simple rig count since many of the productivity gains discussed earlier mean that each rig can do more drilling over the same period of time. In 1970, productivity per foot was about 200 barrels of oil equivalent (BOE, that is, total energy found including oil, gas, and any other petroleum products in the well). In 1980, productivity per foot was about 250 BOE. In 1990 productivity per foot drilled reached about 400 BOE, remaining at these levels or better through 1994.[14]

---

[10]U. S. Department of Energy, Energy Information Administration, *U.S. Crude Oil, Natural Gas and Natural Gas Liquids Reserves 1991 Annual Report*, p. 16; and U.S. Department of Energy, Energy Information Administration, *U.S. Crude Oil, Natural Gas and Natural Gas Liquids Reserves 1994 Annual Report*, p. 10.

[11]U.S. Department of Energy, Energy Information Administration, *U.S. Crude Oil, Natural Gas and Natural Gas Liquids Reserves 1994 Annual Report*, Figures 7–10.

[12]American Petroleum Institute Research Study No. 064, *U.S. Petroleum Supply: History, Prospects and Policy Implications*, September 1992, Table 5 at p. 17. Note, however, that this table aggregates exploration and well-development data.

[13]Ibid.

[14]Ibid. See Figure 30 at p. 16 for success rates and Figure 32 at p. 17 for finding rates. The average success rate masks the changing well composition (ratio of developmental to exploratory wells) in the data. In the absence of changing composition, the data on off-shore and on-shore wells are contradictory. Table 5 reports that the average success rates on-shore went from 55.1 percent in 1970 to 69.3 percent in 1990, and off-shore from 65 percent in 1970 to 42 percent in 1990. But if off-shore drilling

Thus, productivity per foot drilled has effectively doubled in 20 years. But notice how easy it is to cite selectively from the above figures to reach the opposite claim. A proponent of the bursting bubble view of supply might note that the number of active rigs in 1994 is one-third that of 1982. This seems catastrophic until one realizes that the productivity per rig and per foot drilled each doubled in the same period, while inventory requirements were reduced. The surprising fact is not that the rig count dropped after the mid-1980s, but that it did not drop further.[15]

The costs of production over the period fell sharply, reflecting the productivity gains. In nominal dollars per BOE, total exploration, development, and acquisition costs were $0.62 in 1977, rose to a peak of $1.36 by 1985, and then fell to $0.67 in 1990, remaining similarly low into 1994.[16] Adjusted for inflation, costs were much lower in the 1990s than in 1977. In sum, reserve replacement costs

---

moved from being mostly development in 1970 to mostly exploration in 1990, then the average success rate off-shore would fall even though productivity was actually rising. This is the most consistent interpretation of all of the data presented in that study. The Institute notes exactly this interpretation to explain why lower-48 state offshore rates fall while both lower-48 on-shore and Alaskan drilling success rates rise in an earlier Table 28 at p. 15. But precise percentages between exploration and development are also not given in their discussion.

[15]For an example of what misunderstanding the data surrounding rig counts can do to otherwise sane and sober individuals, see Donald Hodel and Robert Deitz, *Crisis in the Oil Patch* (Washington: Regnery Publishing, 1994).

[16]For 1990 and earlier data, see *Oil and Gas Journal*, June 22, 1992, p. 60. The data presented here are converted from the BOE data of the source by dividing by 5.8 million Btu per barrel. The nominal BOE of exploration, development, and acquisition costs before conversion are $3.61 in 1977, $7.90 in 1985, and $3.34 in 1990. The nominal BOE production costs are $1.28 in 1977, $5.18 in 1985, and $3.88 in 1990. These data show current-year production costs, but use five-year average exploration, development, and acquisition costs because that is how the data are reported; in particular, only the production costs are reported on both a current-year and five-year average basis. In "inflation-adjusted constant 1982 dollar" form, the productivity gains discussed here seem even more pronounced. For data after 1990, see Energy Information Administration, *Performance Profiles of Major Energy Producers 1994*, February 1996, pp. 28–29. The U.S. average cost data reported there are $5.11 per barrel (or about $0.88 per Mcf equivalent of natural gas) in 1993 and a lower number, $4.53 per barrel or about $0.78 per Mcf, in 1994. However, the 1994 data are reported as three-year rolling averages, while the 1970 through 1990 data are reported as five-year rolling averages. In either data set, the long-term trend is still decidedly downward.

were about \$9.50 per BOE in 1987, just after partial wellhead deregulation, and are now about half that![17]

Nuclear Follies: The Shaky Case for Fusion (and Fission)

The uranium industry has experienced many of the same changes in costs experienced in the petroleum industry. For example, from 1980 to 1992, the cost of uranium exploration fell from \$4.84 per foot to \$2.25 per foot, a fall of more than 54 percent, while development costs per foot went from \$3.60 to \$2.31, a drop of 39 percent. Productivity measured as output per employment-year increased in this same period by more than 55 percent.[18] In industries that experience productivity gains because of technological advances, the cost of production falls over time. Nuclear energy is no different.

It is, therefore, ironic that supporters of both the uranium (fission) and the fusion parts of the nuclear industry like to use finite resource arguments to promote their market shares, at the expense of petroleum energy. Uranium is certainly a mineral, and if petroleum is a finite resource then so is uranium. Indeed, the EIA is required by law to analyze the "security" of U.S. nuclear energy production essentially based on the availability of domestic uranium at given prices.[19] Certainly, as a matter of national security policy, the analysis of short-term productive capacity at given costs with current technology is a consideration. But uranium is neither less nor more subject to this form of analysis than petroleum.

Therefore, the current campaign by the fusion side of the nuclear industry to enlarge market share is ironic because it uses arguments

[17]Data from sources in previous footnote. Natural gas production costs show a similar pattern. Natural gas production costs were \$0.22 in 1977, \$0.87 in 1985, but \$0.61 in 1990.

[18]Costs for exploration and development; see the Energy Information Administration's *Uranium Industry Annual, 1992*, p. 23. Productivity measured from same source, at Table ES1, p. 16, as the ratio of 14.2 million surface feet drilled in 1981 using 13,676 person-years, compared with 1.1 million surface feet drilled in 1992 using only 682 person-years. The productivity per person-year in 1981 was therefore 1,038 feet per person, increasing to 1,612 feet per person in 1992, an increase in productivity of labor of 55 percent.

[19]See Section 170B of the *Atomic Energy Act of 1954*, as amended by the *Nuclear Regulatory Commission Authorization Act of 1983*. The most recent required report at this writing was U.S. Department of Energy, Energy Information Administration, *Domestic Uranium Mining and Milling Industry 1992, Viability Assessment*, December 1993.

based on finite resource views of petroleum. For example, the *Washington Post* recently ran an op-ed piece by a Princeton physics professor, arguing that "the long quest for clean nuclear power is going better than you think."[20] The article tells us that clean fusion power is near at hand; we just need more federal research funds to reach that goal sooner. As motivation for this necessity, the first few paragraphs of the article argue that "the effects of a finite fossil fuel supply cannot be ignored. . . . The world's known oil supply will be depleted in about 60 years, and natural gas in about a century." There are centuries of coal supply available, the professor writes, but its use would "only aggravate an already precarious ecological balance," for which reason the world has turned to other supplies, including nuclear fission.

This reasoning is silly for two reasons. First, fission power uses uranium, itself as much a mineral resource as coal or petroleum. We use fission power for atomic energy because of often favorable economics of production, not to escape use of "finite" minerals. But second, if the technological promise of fusion power is so near at hand, cheap to produce, and environmentally clean, then capital would surely be forthcoming from the industry itself to develop this source. Indeed, if the facts claimed for fusion power are true, and yet insufficient private capital is provided, the proper policy response is not to seek government funds but to ask whether existing government control may be the cause of limited private interest.

Energy is admittedly a capital-intensive industry. Today's capital is the result of yesterday's accumulated economic rents and profits.[21] But the economic rents of energy production are captured by regulation in other ways. For example, governments take them as increased taxes; energy is more highly taxed than other industries. And regulators reallocate them to other purposes as "costs." The most ironic of these for the present discussion are the excessive labor costs in the nuclear industry, leading to prudence reviews of nuclear power station plant costs by state regulators. It is difficult to believe that the nuclear energy industry has actually benefited from this erratic

[20]Ronald C. Davidson, "Fusion Dreams: Plugging in the Planet," *Washington Post*, August 14, 1994, p. C3.

[21]In economics "rents" or excess profits are returns to capital in excess of the return that the capital could earn if used elsewhere in the economy. It is this feature of investment that attracts capital to more from less productive uses.

pattern of regulation, much less that serious proponents of its future might now actually seek more of the same.

Any law that proposes regulation of energy markets, in the name of conservation of "scarce petroleum," or based on any other scarce resource argument, is misguided. The "scarce resource" problem allegedly solved by such regulation does not exist in the first place and, even if it did, markets would result in far more rational solutions to energy allocation choices than government mandates.[22]

The Decreasing Relevance of Oil Reserves and Replacement Rates

Because well development is now more predictable, the lag time for creating a given level of new production is substantially reduced. The total capital cost invested in new exploration and development is now less for a given level of production. Consequently, a large inventory of fully or partially explored and developed reserves is no longer necessary. If drilling success probabilities are high enough, data that predict where to drill next are cheaper to hold than are the actual reserves themselves. In the new information-intensive petroleum environment, reserves are not "developed" until just before the oil or gas is needed for market.

Measures of oil reserves reflect changes in geologic knowledge and drilling technology.[23] In 1945, the United States current reserve life for oil was about 12 years and for natural gas about 35 years. By 1970, the reserve lives were about 10 years for oil and about 15 years for natural gas. In 1994, the reserve life for oil was about 9

---

[22]For a good sampling of the vast economic literature on resource scarcity, see: Harold Barnett and Chandler Morse, *Scarcity and Growth: The Economics of Natural Resource Availability* (Baltimore: Johns Hopkins University Press, 1963); *Scarcity and Growth Reconsidered*, V. Kerry Smith, ed. (Baltimore: Johns Hopkins University Press, 1979); Orris Herfindahl, *Resource Economics: Selected Works*, David Brooks, ed. (Baltimore: Johns Hopkins University Press, 1974); Richard Gordon, "Conservation and the Theory of Exhaustible Resources," *Canadian Journal of Economics and Political Science*, 32:3 (August, 1966), pp. 319–26; Richard Gordon, "A Reinterpretation of the Pure Theory of Exhaustion," *Journal of Political Economy*, 75:3 (June, 1967), pp. 274–86; and M.A. Adelman, *The Genie Out of the Bottle: World Oil Since 1970* (Cambridge: MIT Press, 1995), pp. 11–39.

[23]Reserve life is commonly measured as the ratio of "proven reserves" divided by the current annual production rate from those reserves. I prefer to describe this measure as "current reserves life," since it measures the service life only of presently available reserves.

years and for natural gas a bit under 10 years.[24] United States proven natural gas reserves dropped from 289.3 quads in 1970 to 187.2 quads in 1990. If one believes that petroleum and natural gas are available only in fixed total amounts and that we have found all there is to find, these figures inspire panic and even legislation.

Given the effect of technology on drilling success rates and well-head production, it is not at all surprising that "proven reserves" of oil would decrease. Current reserve life should fall as the ability to predict successfully where to drill new wells increases, because rational businesses will minimize capital invested in current reserves or partially developed reserves. Once the drilling success rate is high enough, well development is no longer a highly risky gamble but simply a production decision that happens to have a probabilistic component.

A final consideration must be added to this analysis—replacement rate. The replacement rate tells us the relationship between the total amount of new petroleum production potential found in a given year and the amount of petroleum consumed in that same period. The conventional view argues that reserves are running out because they are finite and cannot be replaced. A falling replacement rate for reserves is used as evidence for this hypothesis.

But consider the facts more carefully. If reserve life is falling because it is cheaper to maintain reserves as electronic data than as actual proven reserves in developed wells, then the amount "discovered" each year will be less than the amount produced. The replacement rate will necessarily be less than 100 percent of annual production as a logical consequence of a reduction in developed "proven" reserves. A falling reserve life and a falling replacement rate are two measures of exactly the same phenomenon: the economic adaptation to changing technology that makes maintenance of large quantities of developed reserves unnecessary.

Table 1 reports the annual "replacement rate" of reserves (in barrels of oil equivalents) of energy produced each year for crude oil (line 1), of natural gas[25] (line 2), and of the sum[26] of crude oil, natural gas, and "natural gas liquids" (line 3).[27]

[24]U.S. Department of Energy, Energy Information Administration, U.S. Crude Oil, Natural Gas and Natural Gas Liquids Reserves 1994 Annual Report, Figure 11, p. 14.

[25]The replacement rate for natural gas has been close to 100 percent since 1986. Indeed, the average rate for the four years 1987–1990 is almost exactly 100 percent. This is also the period in which the cost measures presented later in this chapter show close to competitive market conditions.

Notice that the developments in the petroleum industry resemble manufacturing industries adjusting to "just in time" inventory methods. As the ability to move products between stages of production becomes more predictable, the need to maintain large inventories of completed products at earlier stages of production drops radically. As a consequence, "reserves" as well as "proven reserves" are reduced because proven reserves are simply the final stage of natural gas or petroleum "manufacture." The level of inventory required at each stage of production is determined by the overall efficiency of technological knowledge. As that knowledge increases, the level of inventory required at the final stage of production has been reduced and is still falling.

These quantitative changes are often interpreted by the media as cause for panic: "We are running out of oil! Reserve life has dropped by half!"[28] But falling reserve life does not mean that an emergency program is required to force enough drilling to increase reserve life back to a reserve life appropriate to the technology of 1950. With current technology, an aggregate reserve life of only a few years may be most efficient. Instead of being cause for panic, a falling reserve life indicates that the former high reserve life is no longer needed because the reliability of replacement is much better.

---

[26]Another interesting feature occurs in line 10 of the table. Total petroleum energy replacement (including natural gas liquids) is typically closer to full replacement of domestic energy consumed than is crude oil alone. The low rate of petroleum replacement is part of an overall picture of changing product mix in which market shares are shifting among products. Energy markets are shifting away from domestic crude oil (and its byproducts) and to natural gas, natural gas liquids (and their byproducts), and imported crude oil. This is a rational market response to the relative cost and product demand structures. From a purely economic point of view, the shift in market shares among products is not something that needs to be "corrected" by any act of policy.

[27]Natural gas liquids are products, such as propane, found in oil or gas wells that are usually extracted from the producing stream from those wells and sold in their own product markets. These "liquids" are a small but significant part of the total energy and total economic value found in such wells; they are not discussed in this book since they are not subject to utility-type regulation.

[28]A classical case of this form of "analysis" was presented by the forecasting newsletter *NERA Energy Outlook*, written by resource economist Miriam Stewart and published by National Economic Research Associates, Inc., between 1988 and 1996. Among others, the *Wall Street Journal* regularly cited the "forecasts" made by Stewart.

Table 1

ANNUAL CRUDE OIL, NATURAL GAS, AND TOTAL PRODUCT RESERVE REPLACEMENT RATES, BY YEAR

Line No.	1980	1981	1982	1983	1984	1985	1986	1987	1988	1989	1990
1 Crude Oil	94.38%	82.13%	43.78%	92.36%	115.32%	92.30%	45.64%	106.33%	79.89%	81.40%	84.10%
2 Natural Gas	86.17	111.82	97.03	90.22	82.51	72.25	86.09	70.63	129.10	92.85	109.18
3 Total, Oil, Natural Gas, and Natural Gas Liquids	93.16	102.44	75.13	99.95	98.56	88.51	70.92	90.95	107.03	84.97	97.61

(Source: *Oil and Gas Journal*, June 22, 1992, p. 60, and *Oil and Gas Journal's Energy Statistics Sourcebook*, 6th ed., p. 336.)

The (Slowly) Bursting Bubble

It would be unfair to conclude this chapter without recognizing that the relationship between advancing technology and production cost is increasingly accepted by many major petroleum industry institutions. The Gas Research Institute, for example, publishes an annual forecast of the long-term price and supply of petroleum energy and natural gas. In the August 1994 forecast for the period starting in 1995, the Institute noted:

> As a result of the new technology developed over the last 20 years, lower-48 and Canadian gas resources are larger than they were thought to be 20 years ago. In fact, despite the depletion of the gas resource base since 1970, the lower-48 and Canadian potentials in many estimates are larger than they were estimated to be in 1970. For example the [National Petroleum Council] estimated a [total] remaining lower-48 gas resource potential of 906 trillion cubic feet (Tcf) in 1972. Twenty years later, the NPC estimated a lower-48 gas resource potential recoverable *with 1990 technology* of 906 Tcf. ... The NPC estimated that new technology has reduced the underlying real costs of drilling about 3 percent per year since the mid-1960s [emphasis added].[29]

Thus as costs fell, total supply estimates also increased. The amount recoverable with 1990 technology, not including all other recoverable reserves, was similar to the estimated total of all recoverable reserves of two decades earlier, despite the fact that a significant amount had also been consumed in the intervening period.

Publication of such views by a major petroleum industry forecasting service marks a significant departure from the history and policy otherwise surveyed in this book.

Finally, it is worth noting that the underlying view of this chapter—that energy is simply another technological product whose economics are subject to the ordinary market effects of supply and demand—is not new. In 1973, the *Wall Street Journal* published a comparison of the petroleum crisis with the whale oil crisis of the

---

[29]Gas Research Institute, *GRI Baseline Projection of U.S. Energy Supply and Demand*, 1995 ed., August 1994, pp. 27–28.

19th century by a then-unknown academic named Phil Gramm.[30] Professor Gramm documented how the mid-19th century "crisis" in whale oil supply stimulated technological creativity that made petroleum production economically viable.

The only real mystery about the economics of energy production is not about nonexistent "natural" limits to supply, but why the academic world and government have so persistently ignored reality in favor of expanded government regulation. Analysis of petroleum industry data deflates both the rising bubble view of price and the bursting bubble view of supply.

Petroleum and Natural Gas Industries: Infected by Monopoly?

Are petroleum and natural gas production organized as natural monopolies for which some form of regulation may be appropriate? Is the level of competition sufficient to provide consumers with the benefits of a static as well as dynamically efficient market?

Market concentration refers to both the number of competitors in a market and their relative size and consequent ability to influence market quantities and prices. Market concentration may range from a market subject to a pure monopoly in which only one seller completely dominates the market to one characterized by perfect competition in which no single producer is large enough to influence market price.

The Herfindahl-Hirshman Index: Coming Up Empty

Market concentration can be measured by various means. The simplest is the percentage market share held by each participant. This measure, while showing the relative size of each seller, does not reflect the relative importance of market share as a measure of market dominance. Instead, most scholars now believe that a nominal increase in percentage share of the market causes a disproportionate increase in relative market dominance of the largest seller and relative decrease in the market dominance of those with smaller

---

[30]See "The Energy Crisis in Perspective," *Wall Street Journal*, November 30, 1973. Similar views about resource industry generally have been expressed in Jerry Taylor, "The Growing Abundance of Natural Resources," pp. 363–78, in *Market Liberalism: A Paradigm for the 21st Century*, ed. David Boaz and Edward H. Crane (Washington: Cato Institute, 1993).

shares. Said differently, market dominance is a nonlinear function of market share.

An improved measure of market dominance that reflects this non-linearity is the Herfindahl-Hirshman Index, or HHI, so named after the economists who invented the concept. The HHI summarizes the market shares of individual companies based on the *square* of the percentage of the share held by each seller. That is, if a company has a 10 percent market share, then the HHI contribution of that company is 100, 10 squared. If a company has a 1 percent market share, then the HHI contribution of that company is 1, which is also 1 squared. Thus, the HHI measure gives disproportionately greater weight to companies with larger shares, reflecting their presumed disproportionately greater ability to influence the total market.

The summation of the HHIs in a market produces a number between 1 and 10,000 that measures the degree of concentration in a particular market. For example, when there is only one seller, that seller has a 100 percent market share, which when squared gives an HHI of 10,000. In contrast, when there are 10 sellers of equal size, then each seller has a share of 10 percent. The square of each share is 100, and the sum of all 10 of those squared values gives an HHI for the market of 1,000. The higher the HHI total for all companies in a particular market, the more monopolistic the competition.

The HHI index is widely used in regulatory contexts to analyze the competitiveness of markets. For example, the U.S. Department of Justice uses the HHI to analyze corporate mergers. Initially, a fairly low HHI measure (referred to as a "vertical restraints index") of only 1,200 was used to decide which mergers deserved more careful regulatory attention. More recently, mergers that increase the total market HHI to a measure above 1,800 are considered to result in markets that are "highly concentrated" and thus subject to more careful scrutiny. Mergers that result in total HHI measures under 1,000 are considered to result in markets that are "unconcentrated" and thus given relatively little scrutiny.[31] The HHI also is used by the Federal Energy Regulatory Commission (FERC) to analyze the competitiveness of various natural gas producing areas.

---

[31]See "U.S. Department of Justice Guidelines for Vertical Restraints" in Commerce Clearing House *Trade Regulation Reports*, No. 687, Part II, January 30, 1985.

The standard FERC uses as a minimum threshold of concern an HHI above 1,000.[32]

A recent study of the HHI measures of concentration in the petroleum industry was published by the American Petroleum Institute.[33] The study showed that no segment of the industry had an HHI above, or even very near, 1,000. For example, ownership of natural gas reserves had an HHI of 184.5. Natural gas production had an HHI of 126.1, down from a high of 257.7 in 1970.

Those were national measures. Some local natural gas market centers, consisting of producing areas near a particular point on an interstate pipeline, could have higher HHIs. The HHI at seven such market centers was examined by the FERC staff, with resulting indices that ranged from a low of 327 at Guymon, Oklahoma, to a high of 1,774 at Blanco, New Mexico.[34] Thus, while some local markets are more concentrated than the national natural gas market, the picture drawn by HHI measures shows that petroleum production is a generally unconcentrated and therefore presumably competitive market.

Productivity and Market Organization: Another Piece of the Puzzle

An alternative measure of market competitiveness is the comparative productivity of "factors of production" used to manufacture a product. That measure follows directly from the traditional economic maxim, "In a competitive market, price equals marginal cost." Marginal cost is a term used in economics to describe the cost of producing the next unit of output in the most cost-efficient manner possible.

If prices equal marginal cost, certain relationships exist among the costs for the various inputs used to produce outputs. One usually thinks about a price (or cost) of an item as dollars per unit to purchase that item. But one could also think about the inverse of price, which is the amount of item purchased per dollar. This inverse of price, or a similarly computed inverse of cost, measures the productivity of each dollar spent.

---

[32]See footnote 2 in the discussion paper of the Office of Economic Policy, Federal Energy Regulatory Commission, "Importance of Market Centers," August 21, 1991.

[33]American Petroleum Institute Discussion Paper #014R, October 1992.

[34]See Table 1 in Federal Energy Regulatory Commission Office of Economic Policy *Importance of Market Centers* discussion paper, August 21, 1991.

In a competitive market in equilibrium (a term used by economists to describe a state of affairs in which no consumers or firms can better their situation unilaterally through changes in their consumption or production behavior), a special relationship exists among the productivities per dollar of factor costs in the production process, and also between the total factor productivity (the inverse of total marginal cost) and the resulting market price: the productivities of all factors will be equal to each other and the productivity of the total marginal cost will be equal to the productivity of the price. Since productivities per dollar are actually inverses of costs or of prices, this same statement can be made in a different way: in a competitive market in equilibrium (the inverse of) the price equals (the inverse of) the total marginal cost of the good per unit and equals (the inverse of) the marginal cost per unit of output of each factor of production used to make that good.

Let us use two of these measures to examine the competitiveness of the petroleum market. To simplify the analysis, let us assume that there are two factors involved in making petroleum. The first factor would be called "production," defined as the process of extracting something useful from an existing well (whether oil or gas or any of the associated useful chemicals that might also come from oil). The second factor will be called "exploration and development, including acquisition" (or EDA, for short). The term will refer to the process that begins with raw earth and some contracts that permit an "oil" company to use it; adds computer parts, drilling equipment, and some skilled labor; and concludes with the creation of a hole into a pool of oil or gas capable of becoming a producing well.

For the wellhead market to be considered a competitive market in static equilibrium, two conditions must be established. First, the productivity per dollar of production must be shown to be the same as the productivity per dollar of EDA. Note that if the productivity per dollar is the same for the two factors of production, then the cost per unit will also be the same for each factor. Second, the market price will equal the total marginal cost per MMBtu's.[35]

---

[35]If the market is competitive but not in equilibrium, then these conditions should hold, but there will also be "economic" profits or rents. In this analysis I use annual aggregate data for the entire production industry. A year, therefore, is the measure of a marginal unit of time. Certainly, for many business decision purposes this is an appropriate measure. If the year is the unit of time, the second condition becomes: the marginal cost per unit in the year must equal price for the year. The data in

Consider first the conditions regarding marginal productivity of factors of production for the decade of data summarized in Table 2. In a state of competitive equilibrium, those marginal productivities must be equal among all factors of production. Table 2, line 1 measures the productivity per dollar spent on EDA measured as energy equivalent to that contained in a barrel of oil (BOE). Line 2 indicates the productivity per dollar for production of energy measured as energy equivalent to that in a barrel of oil.

During the period 1980 to 1986, the market cannot be considered to be in competitive equilibrium, because the two measures of marginal productivity on lines 1 and 2 of Table 2 differ substantially in that period. But from 1987 through 1990, the two measures of marginal productivities are quite similar. In fact, for all practical purposes, the two measures for that period are the same. Thus, the petroleum market can be considered to be in competitive equilibrium after 1986 by this test.

Another algebraically equivalent method of establishing the existence of competition is specified in terms of cost rather than productivity. Assuming the simplified view of production used here, when the productivities of the two factors are the same, their marginal costs per MMBtu also should be the same.

The facts regarding costs for the factors of production are found in Table 3, lines 1 and 2. Line 1 shows the marginal cost for EDA and line 2 shows the marginal cost for production. As with the information in Table 2, lines 1 and 2, the marginal costs of the factors differ in the period 1980–1986, indicating that the market is not in competitive equilibrium. But for the period 1987–1990, the marginal costs for the factors are very similar. Thus, after 1986 two conditions for a competitive market are met: the marginal factor productivities per dollar expended are approximately equal between the factors,

---

Tables 1, 2, and 3 use aggregated costs for oil and natural gas wells, and thus represent "petroleum" in the more general sense. The data in Table 2 are given in "barrels of oil equivalent" (BOE), the amount of energy typically found in a barrel of oil, about 5.8 MMBtu. Thus in Table 2, data are converted to MMBtu values using the 5.8 conversion factor, since MMBtu's are the basis for other price and cost analysis in this book. Since each year or group of years represents a different marginal period, the two conditions verify the existence of competition only for that period. Separate empirical calculations must be done for each year or group of years to establish the existence of competition in those periods.

Table 2
FACTOR PRODUCTIVITY IN BARRELS OF OIL EQUIVALENT PER DOLLAR OF COST, BY YEAR

Line No.	1980	1981	1982	1983	1984	1985	1986	1987	1988	1989	1990
1 Productivity of EDA Costs (Exploration, development, and acquisition)	0.176	0.130	0.096	0.138	0.138	0.134	0.182	0.294	0.286	0.248	0.290
2 Productivity of Production Costs	0.134	0.161	0.176	0.180	0.199	0.193	0.272	0.289	0.275	0.262	0.257

(Source: *Oil and Gas Journal*, June 22, 1992, p. 60, and *Oil and Gas Journal's Energy Statistics Sourcebook*, 6th ed., p. 336.)
Note: Productivity measured in Barrels of Oil Equivalent energy content, at 5.8 MMBtu per BOE, per dollar of cost.

Table 3

MARGINAL COSTS FOR FACTORS OF PRODUCTION AND TOTAL MARGINAL COST COMPARED WITH PRICE, BY YEAR

Line No.	1980	1981	1982	1983	1984	1985	1986	1987	1988	1989	1990
Nominal Costs, $ per MMBtu											
1 Exploration, development, and acquisition	$0.981	$1.330	$1.789	$1.245	$1.246	$1.290	$0.946	$0.587	$0.602	$0.696	$0.594
2 Production	$0.550	1.069	0.980	0.960	0.866	0.894	0.634	0.596	0.628	0.657	0.670
3 Total, production plus EDA costs	$1.531	2.399	2.769	2.205	2.112	2.184	1.580	1.183	1.230	1.353	1.264
4 Natural gas wellhead price	$1.590	1.980	2.460	2.590	2.660	2.510	1.940	1.670	1.690	1.690	1.720
5 Gross margin	$0.059	(0.419)	(0.309)	0.385	0.548	0.326	0.360	0.487	0.460	0.337	0.456

(Source: *Oil and Gas Journal*, June 22, 1992, p. 60, and *Oil and Gas Journal's Energy Statistics Sourcebook*, 6th ed., p. 336.)

and the marginal costs per MMBtu of the two factors are also approximately equal. (Indeed, these two statements are essentially the same information in different form.)

In competitive markets total marginal cost equals market price, and economic (or excess) profits are zero.[36] In a perfectly competitive market in static equilibrium, line 5 of Table 3 would be zero. Total marginal cost is on line 3, market price is line 4, and line 5 is the difference: the gross margin or gross profit.[37]

Now compare the data in Table 3 to the theory. Except perhaps in 1980, when the gross margin approximates zero, in no year does the total marginal cost equal the price. From 1983 forward the gross margin is between 32 and 50 cents, and between 1987 and 1990 the gross margin seems to stay in the range of about 46 cents. The same patterns continue in both aggregate petroleum market data and separate natural gas market data after 1990.[38] The persistent presence of a surplus above marginal cost is not characteristic of a market in a static competitive equilibrium. It *is* characteristic of a market in a more dynamic state in which profits (positive differences between costs and prices) can be regularly produced. In a dynamic market, companies keep inventing new things that upset the previous cost patterns to produce the source of profits. In the real world, economic

---

[36]Remember that normal or "accounting" profits must be at least the risk-adjusted return elsewhere in the economy.

[37]In this analysis, this difference is probably more correctly described as a pure economic rent. In normal accounting terms, it is the profit.

[38]After the removal of wellhead price controls in 1985, the natural gas market increasingly exhibited the characteristics of a dynamic competitive market. Reserve replacement costs remained steady at a bit over $5.00 per barrel, or about 88 cents per MMBtu through 1993, and then fell in 1994. (See pp. 28–29, in Energy Information Administration's *Performance Profiles of Major Energy Producers* 1994, February 1996.) Both average finding costs for a barrel of oil equivalent. (See Figure ES-3 and pp. xi–xii of U.S. Department of Energy, Energy Information Administration, *Natural Gas 1995: Issues and Trends*, November 1995.) The U.S. domestic reserves replacement rates also remained similar to the 1990 figures. These were about 93 percent for oil in 1991 and advanced steadily to about 98 percent in 1994; while domestic reserves replacement rate for natural gas was about 99 percent for the entire period, and actually a bit over 100 percent in 1994, essentially flat. (See Table 1, p. 4 in Energy Information Administration, *U.S. Crude Oil, Natural Gas, and Natural Gas Liquids Reserves 1994 Annual Report*, October 1995.)

profits are rarely zero. Rather, competitors generate profits by reducing costs through innovations, finding new customers, and producing new products. The petroleum market is consistent with this description.

Inasmuch as the overall measures of the aggregate petroleum (including natural gas and other wellhead products) market since 1986 produce a "stable" outcome (in the sense that the measures are similar from year to year) and also produce a profitable result, the period starting in 1987 may be described as one of "stable dynamic equilibrium," rather than the textbook "static equilibrium," because of the existence of a positive gross margin (that is, a profit or spread between costs and prices, whereas in static equilibrium profits are zero). But it is *exactly* what one expects to find in a healthy competitive market undergoing continuing technological innovation.

Diversity of Company Performance: Monopoly Shrugged

If a market is competitive but dynamic, then one also expects to see diversity in company performance. American Petroleum Institute[39] as well as Department of Energy[40] data suggest the existence of a great diversity of production cost structures among producing companies. These data show that certain firms have a significant competitive edge in exploration over others in the industry.

For example, in 1991, the reported or "sample" companies had a 57.8 percent success rate for finding oil-producing wells while the U.S. average finding success rate was 23 percent. This implies that the finding success-rate average for the remainder of the U.S. industry was 13.6 percent. The success-rate ratios are similar for 1990. The sampled companies were on average more than four times better at finding new natural gas supplies than were the remaining companies. This implies that an astonishingly high cost advantage must accrue to the high success-rate companies.

For development wells, the sample companies had a 92.6 percent success rate in 1991, while the U.S. average was 81.5 percent and the average for the remaining nonsampled companies was 79 percent.

---

[39]See American Petroleum Institute Research Study #064, September 1992.

[40]U.S. Department of Energy, Energy Information Administration, *Performance Profiles of Major Energy Producers 1991*, December 1992, especially Tables B23 and B26.

Though these differences are not as dramatic as those for the exploration success rate, the numbers also show a significant cost advantage for the sampled companies.

The relationship between exploration and development success is also worth noting. In 1985, the sampled companies had an exploration success rate of 39.2 percent, which was lower than their 1991 rate. Their development success rate, however, was 92 percent, similar to that of 1991. Thus, the increase in success for the sampled companies between 1985 and 1991 was essentially all in findings success, which jumped from 39.2 percent to 57.8 percent, an increase to one and a half times the previous level. Inasmuch as development success rate is constant, the relative amount of "proven" reserves is altered because of the change in finding success rate.

Let us translate the percentages into concrete terms. In 1991, the sampled companies accounted for only 19 percent of the total exploration and development wells drilled, or 5,570 wells of a total 29,500. The sampled companies also accounted for 22 percent of the exploration wells, or 924 of a total 4,340. But the sampled companies found 534 successful exploratory wells, and the rest of the industry, drilling about five times as many holes, found only 466 successful wells. The advantage given by high success rates is apparent.

These statistics corroborate the theory that changes in information management are primarily responsible for increasing productivity and reducing costs in the petroleum industry. Producers with better information productivity have much better cost structures. But the data also imply the converse: the petroleum production market may not be as perfectly competitive as measures based entirely on market shares imply because the companies with the lowest costs (arising out of the successful use of geological information technology) may exert a disproportionate influence on the market. Their cost structures give them a relative advantage over their more numerous, smaller, or more costly competitors. But that's as it should be. To the innovative go the profits.

Conclusion

Petroleum, natural gas, and uranium markets have experienced technological innovations that severely undermine the view that supply is finite, shrinking, and increasingly costly. Instead, computer technology combined with increasingly sophisticated geological knowledge has dramatically increased our ability to exploit existing and locate new reserves.

Market share and cost data suggest that petroleum and natural gas markets are competitive and innovative. Consumers are well served by both characteristics, precisely because the latter do produce economic ("excess") profits that go to the most innovative (and, hence, lowest cost) companies.

3. Do Electric and Natural Gas Utilities Need to Be Regulated?

Popular views of the economics of local utilities are as mistaken as popular views of petroleum markets. A fundamental myth of utility economics is that local energy distribution is necessarily a monopoly. This view is false, for very much the same reasons that petroleum production is not a finite resource. Technological change, and especially development in information technology, has completely transformed the local energy industry. Traditional assumptions about which company has the "natural" advantage to perform which service are often the reverse of the present politically assured arrangements. The possibility of greater competition in local utility services is limited far more by government than by economics.

Monopoly 101

The basic myth of energy distribution is the myth of the natural monopoly. A natural monopoly exists if marginal costs become smaller continuously as output increases.[1] In those circumstances a single firm can serve the entire relevant market at a lower cost than can multiple sellers.[2] In the context of the natural gas market, many argue that a single long-haul pipeline can serve its customers at a lower cost than can several competing pipelines between the same source of supply and the same point of market. Similarly, the argument says that a single local natural gas distributor can serve its local distribution market area at a lower cost than can competing

[1]This is not the definition of "natural monopoly," but if this condition holds then certainly natural monopoly also exists. The definition is that average costs of a single firm decline over the entire range of relevant market.

[2]The natural monopoly idea seems to have originated as a philosophical notion in the writings of John Stuart Mill, in an 1846 discourse concerning railroads, an industry no longer regulated. See John Stuart Mill, *Principles of Political Economy*, 1848, reprinted by Colonial Press, London, 1900. For a classic modern reference see Alfred Kahn, *The Economics of Regulation*, vol. 2, ch. 4 (Cambridge, Mass.: MIT Press, 1988).

gas distributors; a single local electric company can produce, transmit, and distribute electricity in a local market area more efficiently than can competing electric companies.

Since such peculiar utility economics are claimed to be the "natural" results of technology, it follows that, left to their own devices, one of the competing utilities will acquire an actual monopoly. One firm will grow sufficiently large to take advantage of economies of scale and scope[3] and soon vanquish the competition, leaving the firm without competitors now and forever. Consumers, accordingly, will have no alternative but to buy from the monopoly provider, leaving them vulnerable to the most dramatic sorts of price gouging and economic victimization that the mind can imagine. Therefore, rather than let unbridled natural monopoly become unfettered actual monopoly, we should face the inevitable and regulate utilities. We can do this fairly, goes the argument, by creating independent regulatory commissions empowered to grant local monopoly franchises in exchange for the right to regulate the prices of the monopolies, thus preventing exploitation. The notion of a regulatory compact is sometimes applied to describe the supposed contract formed between utilities and regulators by the exchange of monopoly territories for price controls.[4]

How "Natural" Is the Electric Utility Monopoly?

If this economic diagnosis of the electricity industry were correct, then we should expect to find evidence of "natural monopoly" in the hazy mists of history before the advent of utility regulation. For a few decades, electric utilities were relatively unrestrained by such

[3]Economies of scale exist if the marginal costs of production fall as the output of a firm increases. Economies of scope exist if the marginal costs of production of commodities fall if they are produced by the same firm rather than by separate firms. A supermarket, for example, exploits the economies of scope that exist in the retail sale under one roof of food, meat, books, and other items that previously were sold in separate stores.

[4]It is important to note that even the textbook electricity monopolist faces some constraints on the degree to which he can "gouge" consumers. Excessive prices will reduce energy consumption and beyond a certain point reduce the monopolist's profits below what a lower price might have brought. Self-generation can replace electricity service from the grid for larger industrial users. Finally, consumers can, and do, move to service regions where electricity costs are lower.

"enlightened" regulatory compacts. What, then, does the historic record show?

We do not need to guess. Gregg Jarrell has compared the condition of electricity markets in 1912, when only 12 states regulated utilities, and in 1917, when 25 states did so. Energy historian Robert L. Bradley Jr. summarized the Jarrell thesis as follows:

> If the natural monopoly model for regulation was correct, the states that first implemented statewide public utility regulation should have been the ones that had the highest rates and the lowest quantity supplied. Conversely, the states with the lowest rates and the highest quantity supplied should be the last to implement regulation.[5]

But Jarrell found:

> Utilities in early-regulated (ER) states had 46 percent lower prices, 38 percent lower gross profit margins, and 23 percent higher output than did utilities in later-regulated (LR) states. By 1917, after state regulation was established in ER states, prices and profits had risen and output had fallen.[6]

In other words, the most competitive, least-cost parts of the industry were regulated first, and the result of regulation was to increase price! As Bradley and Jarrell make clear from detailed analysis of the historical record, this was no accident. Samuel Insull, an ally of Thomas Edison and head of the Chicago Edison Company, lobbied for state regulation as a defense against competitors. Through political associations representing the interests of the industry (today collected as the Edison Electric Institute) Insull successfully lobbied for state creation of utility regulation and then for state protection of utilities. If Insull had argued that the state should protect utilities from competition so they could raise prices, he would not have been politically successful. So instead, Insull argued that utilities were natural monopolies, and that states should guarantee utilities protected territories in return for price regulation.[7] Thus, what really

---

[5]See Robert L. Bradley Jr., *Oil, Gas, and Government: The U.S. Experience*, Vol. II (Lanham, Md.: Rowman & Littlefield, 1996), pp. 1268–74.

[6]G. A. Jarrell, "The Demand for State Regulation of the Electric Utility Industry," *Journal of Law and Economics*, vol. 21, 1978, pp. 292–93.

[7]See Bradley, p. 63. Gabriel Kolko makes a similar argument about the regulation of railroads by the Interstate Commerce Commission in *Regulation and Railroads* (Princeton: Princeton University Press, 1965).

happened in the proliferation of state utility commissions in the first third of this century was that Samuel Insull and his political allies, the Progressives, intentionally and successfully undermined a competitive market in electricity.[8]

Utility Costs: Not Consistent with Monopoly

What about today's utility companies? An examination of utility costs casts serious doubts upon the diagnosis of natural monopoly in the industry. The main reason is that capital costs have relatively little to do with the price of electricity or gas; the cost of electricity is driven instead largely by managerial practice and, surprisingly enough, taxes. As we shall see, this is entirely inconsistent with what one might expect of a monopoly industry. Even if capital costs of a modern utility exhibit economies of scale, capital costs are such a small fraction of marginal costs that *total* marginal costs do not decrease with firm size.

The average delivered cost of electricity in 1994 was about 6.9 cents per kilowatt-hour. Figure 1 reveals how those costs are broken down.

Interestingly, the price of fuel is less than a third of total cost of electricity; management, information, and accounting costs are almost as important to the average utility's bottom line. Taxes are a larger corporate expense than the costs of capital recovery.

Not only are taxes one of the largest costs of utilities, they are a disproportionately large part of utility costs compared with the share

[8]For a detailed overview of the breathtaking cynicism, cold calculation, and dishonest machinations of Insull and his allies in this campaign, see Marvin Olasky, *Corporate Public Relations: A New Historical Perspective* (Hillsdale, N.J.: Lawrence Erlbaum Associates), 1987, pp. 33–43.

[9]Data on components of cost for electricity are from U.S. Department of Energy, Energy Information Administration, *Electric Power Annual 1994*, Vol. II, November 1995, pp. 15, 34. Costs derived as revenues per kilowatt hour from Table 1, distributed by cost as percentage of revenue from Table 11. Fuel costs by fuel types are estimated from data on fuel generation percentages by fuel type derived from Table 1, applied to kilowatt hour costs. This method probably slightly overestimates the costs from coal and underestimates the costs from petroleum and natural gas. See Figure 1, p. 37.

[10]Percentages derived from Table 8, "Composite Statement of Income for Major U.S. Investor-Owned Electric Utilities," p. 30 in *Electric Power Annual 1994*, vol. II (Operational and Financial Data), Energy Information Administration. Computed by using the electric utility cost and revenue data, converted to percentages, and applied to the 6.9 cent average customer kWh cost for the year. Percentages therefore reflect "typical" percentages of all U.S. electric utility costs, as desired by EIA. See Figure 1, p. 37.

Figure 1
AVERAGE ELECTRIC UTILITY COSTS PER KILOWATT-HOUR

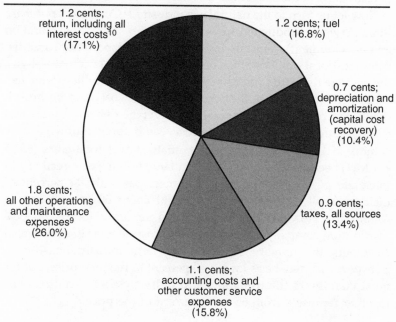

1.2 cents; return, including all interest costs[10] (17.1%)

1.2 cents; fuel (16.8%)

0.7 cents; depreciation and amortization (capital cost recovery) (10.4%)

1.8 cents; all other operations and maintenance expenses[9] (26.0%)

0.9 cents; taxes, all sources (13.4%)

1.1 cents; accounting costs and other customer service expenses (15.8%)

of taxes in costs of other industry. A 1986 study used three decades of Internal Revenue Service data to compare taxes as a percentage of total industry costs and taxes to a percentage of utility costs.[11] The study used data on all utilities, not just electric utilities. The study found that in 1960, the public utility sector paid 11.8 percent of gross revenues as taxes, while the average of all industries was 5.1 percent of revenues paid as taxes. In 1970 utilities paid 9.7 percent of revenues as taxes, while all industry paid an average of only 4.7 percent of revenues as taxes. In 1980, the percentages were 6.6 percent for public utilities and 4.2 percent for all industry. Thus,

[11]William K. Kafoglis, "Tax Policy and Public Utility Regulation," in John C. Moorhouse, ed., *Electric Power, Deregulation and the Public Interest* (San Francisco: Pacific Research Institute for Public Policy, 1986), p. 100.

throughout more than two decades public utilities paid at least 50 percent more of their revenues as taxes than did other industries.[12]

Now let us turn to the natural gas industry.[13] Customers purchase three services: production (including for this purpose exploration and development), transmission (long haul transportation) and distribution (local delivery). For residential customers of natural gas, the average delivered price in 1994 was $6.41 per million cubic feet (Mcf) delivered. Figure 2 shows how those costs can be broken down. For commercial customers, the average delivered price was $5.43. Figure 3 shows how those costs can be broken down.

Industrial customers have substantially lower total costs ($3.05 per Mcf) because they take service in large volume or directly from interstate pipelines. Industrial customers pay only 40 percent of their total bill ($1.22) for transport and about 60 percent for the commodity. However, smaller industrial customers located on local distribution company systems may pay costs similar to commercial customers. In summary, for some very large industrial customers transport cost may be as low as 40 percent of the total price, but for most customers the transport and delivery cost of natural gas together represent from 65 to 70 percent of the price.[14]

---

[12]The appearance that the percentage of revenues paid as taxes by utilities is converging toward that for all industries is probably deceptive. In that same period, energy cost as a percentage of total utility cost may have increased disproportionately because of the way in which the costs were treated by regulators, while taxes as a percentage of other costs were probably constant or increasing. Since the price for energy was progressively more market-determined, and also lower, over the 1980s, it would be interesting to see how the ratios would compare using data for 1990. If my speculation here is correct, then the spread between taxes as a percent of utility revenue and taxes as a percent of all industry revenue should have increased from 1980 to 1990.

[13]Cost comparisons for natural gas utilities are more complicated than for electric utilities because natural gas companies are generally not vertically integrated. Production, transmission, and local distribution are performed commonly by separate companies because of provisions in the Public Utility Holding Company Act and Natural Gas Policy Act. The current structure of the natural gas industry is well documented in a publication of the U.S. Department of Energy, Energy Information Administration, *Natural Gas Annual 1991, Supplement: Company Profiles.*

[14]Data on components of cost of natural gas are derived from industry averages reported by the U.S. Department of Energy, Energy Information Administration, *Energy Policy Act Transportation Study: Interim Report on Natural Gas Flows and Rates,*

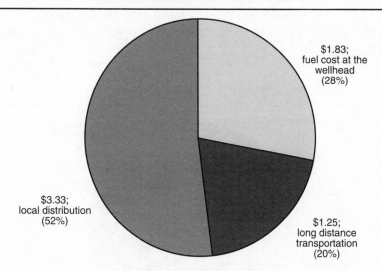

Figure 2
RESIDENTIAL NATURAL GAS COSTS

$1.83;
fuel cost at the
wellhead
(28%)

$3.33;
local distribution
(52%)

$1.25;
long distance
transportation
(20%)

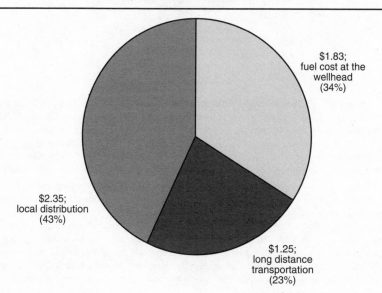

Figure 3
COMMERCIAL NATURAL GAS COSTS

$1.83;
fuel cost at the
wellhead
(34%)

$2.35;
local distribution
(43%)

$1.25;
long distance
transportation
(23%)

Most costs of a natural gas transmission or distribution company are not in physical plant expenses. Consider Figure 4, which breaks down the average costs of long-distance natural gas delivery.

Similar expenditures exist for the local distribution companies (Figure 5).

Figure 4
AVERAGE NATURAL GAS TRANSMISSION COSTS

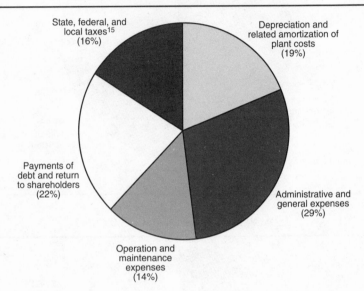

State, federal, and local taxes[15]
(16%)

Depreciation and related amortization of plant costs
(19%)

Payments of debt and return to shareholders
(22%)

Administrative and general expenses
(29%)

Operation and maintenance expenses
(14%)

October 1995, pp. 56, 59. The computation assumes that all classes pay the same average cost for a commodity, which is likely true for customers purchasing from distribution companies but may misstate commodity costs, especially for larger industrial customers able to contract for their own supply. Figure 21 (p. 59) reports a total distribution and transmission cost for industrial customers, at $1.22, which is lower than the $1.25 shown on that same Figure 21 for the transmission component alone.

[15]See U.S. Department of Energy, Energy Information Administration, *Natural Gas 1992: Issues and Trends*, March 1993. For transmission companies, the percentages given above are derived from p. 60 data on fixed transmission costs of a "composite" transmission company that summarizes actual data filed by all interstate pipelines, using 1991 costs. In addition to the costs cited, about 5 percent of costs would typically be expended for "variable" costs, which is essentially compressor fuel. Since this essentially is a fuel cost, I omit it from this computation. This cost is, however, related to the energy "losses" involved with natural gas transmission, a measure of efficiency discussed elsewhere in this book.

Figure 5
AVERAGE NATURAL GAS DISTRIBUTION COSTS

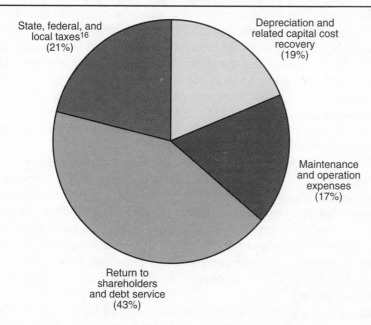

State, federal, and
local taxes[16]
(21%)

Depreciation and
related capital cost
recovery
(19%)

Maintenance
and operation
expenses
(17%)

Return to
shareholders
and debt service
(43%)

The cost structures of electric and natural gas utilities are consistent with neither traditional engineering assumptions nor the decreasing marginal cost perspective of industrial organization economists. The traditional engineering perspective argues that natural

---

Percentages are derived from the local distribution cost structure of a typical local distribution company presented on p. 79. The cost probably overstates the cost of return and interest, compared with more current interest rates. It also probably understates the proportion of costs for administration. However, my practice throughout this book is to take nominal data from self-evidently reliable sources. Here as elsewhere, correcting the data by performing even only the most obvious technical adjustments would generally make my case stronger.

[16]Ibid. For distribution company expenses, percentages are derived from the local distribution cost structure of a typical local distribution company presented on p. 79. The cost probably overstates the cost of return and interest, compared with more current interest rates. It also probably understates the proportion of costs for administration. However, my practice throughout this book is to take nominal data from self-evidently reliable sources. Here as elsewhere, correcting the data by performing even only the most obvious technical adjustments would generally make my case stronger.

gas distribution is a capital-intensive local natural monopoly because local distribution of natural gas is done through pipes.

In pipes, there is a geometric relationship between the diameter of a pipe and the volume that can flow through that pipe, but there is only a linear relationship between diameter and the amount of material that constitutes the pipe. Therefore, smallish increases in pipe diameter yield disproportionately large increases in delivery capacity. It is "obvious" that only one company with a proportionately larger single pipe should have the right to distribute gas locally, because such a single company will necessarily be more efficient than any group of competitors might be.[17]

If the premises of the engineering view were true, that engineering economics drive total marginal cost, one would expect company costs to be largely directed in the amortization of the cost of the plant—that is, in depreciation. In fact, the preceding summary shows that as much or more of the total cost to customers goes for taxes as for capital cost recovery by the utilities. Various forms of administration represent a larger percentage of costs than either taxes or capital recovery. Capital recovery (depreciation) is less than 20 percent of costs for all utilities.

In addition, utility costs are not characterized by the economies of scale that we expect to find in a natural monopoly. We would expect the data to show that the larger the utility, the lower its distribution costs. Empirical studies, however, show a statistically insignificant relationship between average cost and number of customers. Costs are at their minimum in the first 10 percent of the range of output (measured as number of customers).[18]

Two (or More) Grids: Better Than One?

One of the central tenets of regulatory belief is that, even if the generation of power is more competitive than was once believed, the electricity *transmission grid* remains a natural monopoly. Given that the grid is little more than wood, wire, steel towers and a dispatch center, two, three, four, or more grids could operate in a

---

[17]Similar reasoning is used in long-haul transmission of natural gas, as well as electricity generation and distribution.

[18]See Asghar Zardkoohi, "Competition in the Production of Electricity," in John Moorhouse, 1986, cited in previous notes, pp. 83–84.

given service territory. Indeed, economist Vernon Smith has observed that "the existence of multiple parallel transmission and generation units owned by the same firm makes it plain that the traditional static argument for natural monopoly is contradicted by observation."[19]

A more dynamic argument against regulation would recognize that customers might lose from the loss of some of the economies of scale that would exist with a monopoly supplier, but would also argue that customers would gain over time from the innovations that would arise from competition. The tradeoff has been succinctly summarized by Richard Geddes: "The crucial question for the assessment of costs is whether the scale benefits of having a single firm serve the market outweigh the efficiency losses because of the lack of competition."[20]

The evidence suggests that dynamic competitive efficiencies far outweigh scale economies. Competition, even when scale economies may be present, saves money for consumers. A 1981 review of the 23 American cities that were then served by competing electric utilities showed that rates were from 15 to 20 percent lower than for comparable cities with monopoly suppliers. New generating capacity in these cities was constructed at less than half of the cost incurred by monopoly utilities for similar capacity.[21]

---

[19]Economist Vernon L. Smith has studied these comparative issues in more detail and summarized the successful experiences of Chile, Argentina, Great Britain and, especially, New Zealand in opening electric markets to competition. This book does not explore those examples in detail, since they also represent either gains from privatization of government-owned utilities, or markets that are still highly structured by government controls, and usually both. However, if significant economic gains are possible even with such a comparatively high degree of government structuring, then more gains should be possible by simply removing government control entirely. For discussions of these other examples, especially New Zealand, see Vernon L. Smith, "Regulatory Reform in the Electric Power Industry," unpublished paper, January 1995; "Currents of Competition in Electricity Markets," pp. 23–29, in *Regulation*, 1987, no. 2; "Can Electric Power, a 'Natural Monopoly,' Be Deregulated?" in *Making National Energy Policy*, Hans S. Landsberg, ed. (Washington: Resources for the Future, 1991), pp. 131–51.

[20]R. Richard Geddes, "A Historical Perspective on Electric Utility Regulation," *Regulation*, vol. 15, no. 1, Winter 1992, p. 81.

[21]Jan Bellamy, "2 Utilities Are Better Than One," *Reason*, October 1981, pp. 23–30.

Multiple Rights of Way: The Disappearing Reality of "Grid-Lock"

Existing monopoly transmission grids can be circumvented by using rights of way currently dedicated to other purposes. For example, the telecommunication grid represents one alternative means to deliver electricity. In fact, most city lots and many rural lots are connected to as many as six such right-of-way grids: electricity, natural gas, sewer, water, telephone and cable. All locations indirectly or directly are connected to the street and highway networks and to oil and natural gas pipeline networks. Also, public or private railroad lines could serve just as well for power transmission grid rights of way; Amtrak, for example, might serve as a right of way for electric power, natural gas, or any other long-haul transmission grid requiring surface connections. Many other state or municipally owned rights of way also could be opened for transmission use by competing entrants to power markets. And since even government taking of rights of way requires compensation, private construction of rights of way in return for compensation to landowners is also feasible.

The information management skills of existing utilities are useful in what we think of as the information and communications industry. One of the more successful new entrants to the telephone industry is WilTel. WilTel was a spinoff company from the Williams Companies management group, which uses techniques and ideas first honed by Williams in operation of oil and natural gas pipelines. After establishing itself in the telecommunications industry, it was sold as a separate company in 1994 for $2.5 billion.[22]

Nor is the much hyped "information highway" immune from invasion by energy companies accustomed to physical transport problems. Entergy, a major electricity company located primarily in the southern United States, has announced its intention to enter the local integrated information businesses, more commonly associated with video, telephone, or cable television distributors. This should not surprise: information distribution businesses require a combination of management skills already familiar to many kinds of energy companies and may also require distribution facilities including

---

[22]WilTel was started in the late 1980s by personnel from the Williams Companies management group. See *Oil and Gas Journal*, December 19, 1994, pp. 27–28.

rights of way already owned by energy distribution and transmission companies. Congress, and perhaps some of the present telecommunications companies whose markets are protected by present law, was so disturbed by the prospect of energy utility companies competing with telephone and cable television operators that the U.S. House of Representatives held hearings on that subject in 1994.[23] At those hearings existing energy utilities described their communications networks. For example, Paul Denicola, the president of Southern Company, which provides electricity services throughout the southeastern United States, described his company's system:

> Like most utility companies, the Southern Company has a significant and extensive telecommunications system which it uses primarily to operate our extensive electric system and provide internal communications. The Southern Company's facilities include approximately 1,700 miles of fiber optic cable with several hundred more planned. In addition, Southern has extensive analog and digital microwave networks and several wireless radio dispatch systems. We are already looking . . . for permission to establish prototype energy management systems using fiber/coaxial cable.[24]

The possibility of electric power companies successfully entering the telecommunications business is not speculation. Although the purpose of the electricity transmission grid is to control electric devices, once a customer is connected to the grid, other services can be provided. This has been demonstrated by the municipal system of Glasgow, Kentucky, which installed fiber/coaxial cable systems into homes for management of customer load. But the system evolved, so that

> Glaswegians now have cable television channels courtesy of the Plant Board, which delivers not only better reception but also direct lines to local school classrooms and a homework channel. . . . In addition, the Plant Board is experimenting with delivering telephone service to its customers.[25]

---

[23]Joint Oversight hearing with the Energy and Power Subcommittee on Registered Holding Company Entry into Telecommunications, July 29, 1994.

[24]Ibid.

[25]Cited in Philip R. O'Connor and Marc A. Vallen, "Information: The Coin of the Realm in a Competitive Electricity Market," *The Electricity Journal*, vol. 7, no. 1, February 1994, p. 22.

Clearly, if electricity companies are capable of using their existing grid to deliver telecommunication services, then cable and telephone companies and well water, sewer, and natural gas companies are likely just as capable of providing electricity service over their rights of way. The right of way is the most difficult part of the system to procure. The rest is wires and various electric devices attached to the wires (and available on the open market); a dispatch center (that such companies already are skilled at operating for other purposes); poles (that cable or telephone companies may already have in place and could share with electric use) or holes (that any of the companies but especially water and sewer companies certainly have); and billing systems (that all of the companies already have). The critical missing component is whether the company has the management skills to make the business work. Not even the most strident advocates of utility regulation have claimed that those skills are natural monopolies that should be regulated.

Thus, it is not beyond contemplation that, in less than a decade, people will have the choice of purchasing energy from what used to be their phone company, television and cable services from what used to be their electric company, electricity from what used to be the gas company, and telephone services from what used to be an oil pipeline, as the WilTel example proves. Indeed, merely counting the possible number of service trucks arriving at the front door tells us that "natural monopoly" does not describe local utilities. The established rights of way for each of these services can be circumvented by adapting any one of the alternative rights of way dedicated to another service.

Natural Gas Utilities: The Power Company's Utilities Challenge?

The distance between a consumer and the source of his energy is far greater when he's igniting a gas burner than when he's flipping a light switch. The difference is due to the physics of natural gas and electricity transmission and distribution. Those differences are not only of great engineering import but are pregnant with economic importance as well. Indeed, natural gas companies might well be able to enter the electricity business and underprice the monopoly power companies—that is, were it not a criminal act for them to do so.

46

Transmission Losses of Energy

Losses occur in natural gas transmission largely because some of the gas itself is usually used to run compressor stations on the pipeline. This use is typically about one-half of 1 percent per 100 miles hauled. Thus, a long-haul carriage of natural gas can cost in total perhaps 2 to 4 percent of the volume of gas carried. Local system losses typically average somewhat less, perhaps only 2 percent of volume.

For electricity transmission, the percentage loss is much higher and typically more literally a "loss." Electric energy is dissipated due to resistance of the wires used to transmit and due to the radiation (magnetic fields) caused by the electric properties of the current flow. Loss percentages of 5 to 10 percent in local distribution (depending principally on distribution voltage levels) and well over 10 percent for transmission of only a few hundred miles are inherent properties of alternating current electric systems.

The tradeoff between local and distant production dictates how natural gas companies meet their peak demands during the winter heating season. In general, producing fields are so far from delivery markets—and long-haul pipes are so expensive relative to storage areas near markets—that the cheapest way to meet peak loads is to build pipelines only so large that they can carry their average daily load. Peak day capacity is generally met by using larger storage fields located near market areas.

Natural Gas: The "Other" Electric Utility?

The clear implication of the tradeoffs between fuel and transmission costs is the possibility that natural gas companies might actually be able to produce electricity at lower cost than electric utilities.

Consider a typical coal-fired electric power plant operating at 10,000 Btu's per kWh with a 10 percent line loss over long-haul power lines.[26] Compare this to a gas-fired plant operating at 8,000 Btu's per kWh and located very close to the retail electricity market with only a 2 percent total line loss on longer haul gas lines and local mainlines. Simple math indicates that the natural gas company

---

[26]The Btu rate of fuel varies by plant and can be different at various times for the same plant. A fuel's "heat-rate" can be affected by the age and design of the plant, operation at close to minimum or maximum possible output, and weather.

could generate electricity to local markets at nearly *28 percent lower* energy costs than electric companies (20 percent better efficiency of energy production plus 8 percent lower losses in transmission). Alternatively, similar costs of delivered electricity could occur even if the price of natural gas is up to 28 percent higher than the cost of fuels used by the electric company, whether that company uses oil, coal, hydro, nuclear or any mix of fuels, including some portion of natural gas.

Natural gas companies are not allowed to compete directly with electric power utilities because the Federal Public Utilities Holding Company Act (PUHCA), as well as traditional local utility regulation, prevents it. Companies are typically granted local distribution monopolies for either gas or electricity. Competitive entry from other companies, even other utilities, is prohibited. Although gas-fired power plants are now the "facilities of choice" for utilities planning new capacity, natural gas companies are still not permitted in most jurisdictions to sell electricity in competition with the electric company.

If the law did not prohibit competition between electric and gas companies, it is entirely possible that natural gas companies could sell electricity at a lower price than could many traditional power companies. And since natural gas companies already have direct access to many consumers, such competition could come sooner than most people think—that is, if they were allowed to enter the market. But the outcome of a fight between natural gas companies and electric utilities for consumers would result from the differences in line losses and in efficiency of transmission and use of generation fuel, not from marginal costs that decreased with output because of large fixed capital costs.

Market Contestability: The Ghost in the Monopoly Machine

Let us assume for purposes of discussion that electric and gas utilities are not natural monopolies but that the number of actual competitors in any market (if deregulation occurred) would be less than five. Would consumers receive the benefits of competition even though the actual number of firms was small? One of the more interesting theories to arise in economics in recent years is the idea that the number of suppliers needed to avoid the effect of monopoly in a market might be very small—perhaps only two, and in some views, one—which is to say, the threat of entry alone may be suffi-cient to constrain firms with "apparent" market power from rais-ing prices.

This line of argument began with the famous paper of economist Harold Demsetz, "Why Regulate Utilities?" cited in the preface to this book. Demsetz argued that even in the case of natural monopoly and only one supplier, the price can be kept to the market price simply by periodic (say, each five years) reauction of the right for local supply. That argument was then expanded in an equally famous work into the idea of analysis of monopoly as "contestable markets."[27] The argument is that a market will provide consumers with the benefits of competition (prices close to marginal costs) even if only one or two producers actually exist, as long as the threat of entry is real. Firms will price near marginal cost because if entry is costless and prices are higher than marginal costs, the higher prices will induce firms to enter the market, thus eliminating any benefits to incumbent firms from their excessive prices.

The theory has been tested in carefully constructed experiments involving only two firms:

> The most significant conclusion of this research is that the behavioral predictions of the contestable market hypothesis are fundamentally correct. It is simply not true that monopoly pricing is a "natural" result of a market merely because firms in the market exhibit decreasing costs and demand is sufficient to support no more than a single firm.

There is clear evidence not only that monopolies contesting to maintain their position exhibit behavior more competitive than theo-retical monopoly predictions, but also that they actually perform up to the standards of the competitive model.[28]

In sum, even if no challenger arose to challenge an incumbent power in a given market, as long as entry was open and economically

---

[27]See W. Baumol, J. Panzer, and R. Willig, *Contestable Markets and the Theory of Industry Structure* (San Diego: Harcourt Brace Jovanovich, 1982).

[28]Dan Coursey, R. Mark Isaac, and Vernon L. Smith, "Natural Monopoly and Contested Markets: Some Experimental Results," *The Journal of Law and Economics*, April 1984, pp. 91–113.

feasible, the incumbent monopolist would likely conduct his business and price his goods and services as if competition were a present fact.

Is Regulation Necessary to Realize "Energy Efficiency"?

For decades, the stated rationale for utility regulation in the United States was the need to prevent exploitation by monopoly distributors and keep prices low. The Energy Policy Act of 1992 ("EPACT" in legislative jargon) provides a different rationale for regulation, in that it requires the retail price of energy to be raised rather than lowered. EPACT postulates that consumers need to face *higher* prices than they would face if utilities were unregulated because of the environmental externalities created by electricity production. The principal tool for implementing this policy is "integrated resource planning."[29] In integrated resource planning, the regulator requires the regulated company to file periodically a comprehensive long- and short-term plan with the regulator. The plan must compare all possible energy supply options with all possible energy conservation options, and then propose to use only the least cost combination. Regulators can modify proposed plans and force companies to implement the modifications.

Integrated resource planning is motivated by the belief that energy markets are characterized by two types of market failure or externalities: the failure of consumers to invest in capital improvements whose cost is less than the savings[30] that result from reduced energy use, and the failure of energy prices to reflect the cost of environmental damages. I will discuss them in order.

Supporters of integrated resource planning believe that consumers do not practice cost-effective energy consumption behavior because of an inability to compare current capital expenditures with savings in future operating costs. Many energy analysts argue that the return on capital implicitly required by consumers through their appliance purchase behavior[31] is often much higher than returns available on

---

[29]Other terms synonymous with integrated resource planning are "demand-side management" and "least-cost planning." This book refers to all of these as "integrated resource planning."

[30]Technically, the present value of the flow of future energy cost savings.

[31]Because they choose appliances that have lower initial costs but higher operating (energy) costs rather than appliances with higher initial costs but lower operating costs.

other assets and thus difficult to reconcile with rational behavior. Fluorescent lights, for example, last longer and use much less energy than incandescent light bulbs, but consumers incorrectly perceive the $20 cost of the former to be "too much" to pay for a light bulb. Integrated resource planning is proposed as a solution to this apparent market failure. But empirical evidence of actual consumer behavior shows consumers are much more rational than this fantasy asserts. Dreyfus and Viscusi estimate the willingness of consumers to spend more now to save energy costs later and find pure rates of time preference on the order of 11–17 percent, rates that are perfectly consistent with returns on other assets.[32] In short, there is no energy consumer market failure.

Even if consumers do have cognitive limitations, the evidence does not suggest that "demand-side management" plans, which are part of integrated resource planning, provide a cost-effective supplement to market forces. For example, the Illinois Commerce Commission studied the complete cycle of all costs for natural gas demand-side management and integrated planning programs from their inception in 1985 until 1994. No program showed benefits greater than costs, and most programs showed that benefits were only about 25 percent of costs (as a result the state legislature repealed the program).[33]

The Energy Information Administration examined the benefits and costs of demand side management nationwide.[34] In 1994 the total cost of demand-side programs was $2.72 billion. The total kilowatt-hours (kWh) saved by the programs was 52,483 million. Thus, the cost per kWh saved was 5.17 cents. In the same year, the sum of fuel, operating and maintenance expenses for nuclear energy was 2.25 cents, and for fossil-fired steam was 2.28 cents. Of these the fuel-only component was 0.61 cent and 1.75 cents, respectively. Thus, in 1994 the country spent $2.7 billion to "save" energy in mandated "efficiency" programs at a cost more than twice that of generating

[32]Mark K. Dreyfus and W. Kip Viscusi, "Rates of Time Preference and Consumer Valuations of Automobile Safety and Fuel Efficiency," *Journal of Law and Economics* 38, 1995, pp. 79–105.

[33]Personal communication from Illinois Commerce Commission Commissioner Ruth K. Kretschmer, 1995.

[34]From Energy Information Administration, *Electric Power Annual 1994*, vol. II, November 1995, pp. 35, 82–83.

that energy directly, and at four to six times the value of the fuel avoided. And this accounting fails even to consider the opportunity costs borne by consumers under such programs, which surely are greater than zero.

Clearly the Illinois results were not anomalous; only the intellectual honesty of the Illinois legislature in correcting a mistake was unique.

Now consider the (false) possibility that regulation has succeeded in its traditional nominal economic objective, to constrain the retail price below what unregulated monopoly would charge. If energy conservation is an objective of integrated resource planning, then the presumed success of traditional regulation to keep price below the monopoly rate suggests one way to induce more conservation immediately: remove retail price regulation, allow retail price to rise to unregulated levels, and let retail home improvement companies compete to aid consumers in saving energy. The results should be similar to those advocated for demand-side energy conservation programs except the cost of regulation would be removed.

Thus, the desired efficiency results of integrated planning could be achieved even in the presence of true natural monopoly by removing all traditional regulation, including the requirement that integrated planning be undertaken. Such deregulation would have the further benefit of restraining price to "only" what an unregulated monopoly would charge, whereas under regulation the cost of regulation is added to the monopoly price levels.

Environmental Externalities: The Regulator's New Clothes?

As the traditional rationales for utility regulation have been criticized, a new rationale has become popular in regulatory circles. The new view claims to look at "true societal marginal costs," also known as "externalities." According to the new approach, utility prices do not reflect "true" resource costs because they do not reflect costs to the environment that are the result of energy use, but external to the utility. This view assumes that tort remedies and existing regulation by the Environmental Protection Agency are inadequate.

The desire to incorporate "externalities" as additional "new" costs in regulatory proceedings falsely assumes that energy companies ignore the "externalities" that their operations create. Centuries of legal history deal directly with the liability faced by parties that harm the property or persons of others.[35] The common law of Britain and the United States especially recognizes the ideas of "trespass" and "nuisance," which occur when an activity of one person (such as to create noxious odors, or smoke, or poisonous products) affects the property or other rights of another person. The law protects the actually or potentially harmed person by devices such as injunctions against offensive behavior and payments of damages for actual harms done. This role of the common law is well known and widely discussed in any first-year law school class on torts. For example, consider this passage from the commonly cited text on tort law of William L. Prosser: "Any physical entry upon the surface of the land is a trespass, whether the entry is a walking upon it, flooding it with water, casting objects upon it, or otherwise" including entry via the air of offensive particles.[36]

A similar result holds for the law of nuisance and other aspects of tort and even contract law. Apparently the only people who are unaware that the law has for centuries treated the problem of externalities with a system of preventive measures, and monetary payments, and other remedies for actual harms are energy utility regulators and their academic advisers. Any company with a competent attorney already knows of the compensable risks of externalities and avoids them.

The environmental "cost" of fuel sources is also at least roughly reflected by the regulatory costs imposed on business by the regulation at the point of fuel use. Those regulatory costs are reflected both directly in the price of the fuel (the cost of fuel production reflects myriad regulatory burdens) and indirectly in the environmental compliance costs for the power facility in question.

Some academics and governments, however, believe that additional action is required. The Massachusetts Department of Public

---

[35]See generally Morton J. Horwitz, *The Transformation of American Law: 1780–1860* (Cambridge: Harvard University Press, 1977), particularly pp. 63–139, and Ronald Coase, "The Problem of Social Cost," *Journal of Law & Economics*, Vol. III, October 1960, pp. 1–44. See also: Bruce Yandle, *The Political Limits of Environmental Regulation: Tracking the Unicorn* (New York City: Quorum Books, 1989), pp. 41–63; Elizabeth Brubaker, *Property Rights in the Defense of Nature* (Toronto: Earthscan Publications Limited, 1995); and Robert L. Bradley Jr., *Oil, Gas & Government: The U.S. Experience,* Vol. II (Lanham, MD: Rowman & Littlefield, 1996), pp. 1268–74.

[36]*Prosser and Keeton on Torts*, 5th ed., W. Page Keeton et al., eds. (St. Paul, Minn.: West Publishing, 1984), especially pp. 70–71.

Utilities provided a graphic demonstration of this belief in a November 10, 1992, order on environmental "externality" values to be used for resource planning by electric companies. The department found that the environmental costs per kWh were *$3.88* for coal and *$1.27* for natural gas. The dollar amounts are emphasized to show that the values are not typographical errors. Typical actual electric bills are around *6 cents* per kWh, using the same sources as the federal Department of Energy data cited in Chapter 1. The Massachusetts Supreme Court eventually found that the department could not consider costs that do not affect the utility's own costs in setting rates, but the Attorney General's office of that state still believes the utility department can consider "externalities," though at present it does not.[37] Those regulatory standards for "least cost planning" costs are from 20 to 60 times greater than the average utilities' total direct costs and perhaps 100 times greater than the respective fuel costs (marginal energy costs) for electrical generation.

Are the numbers reasonable? Serious people believe not. In a similar circumstance, the Executive Officer of the California Air Resources Quality Board advised the California Energy Commission regarding proposed methods for valuing air emissions:

> We believe no one is able to attach credibility to the values for use in electricity resource planning or any other purpose. . . . The uncertainties associated with the use of these numbers are so large as to render the values of questionable utility for the intended purpose, no matter how narrowly defined that purpose might be.[38]

Such estimates have also been criticized as ignoring the economic meaning of costs, which is to say that in economic analysis cost is always considered as a measurement among alternative choices, the "opportunity cost":

> In other words, the economically relevant damage costs relate to the forgone satisfaction from choices that people would

---

[37]General Accounting Office Report to the Ranking Minority Member, Committee on Science, U.S. House of Representatives, *Electricity Supply, Consideration of Environmental Costs in Selecting Fuel Resources*, May 1995, p. 16.

[38]Testimony from James D. Boyd, Executive Director, California Air Resources Board, to the California Public Energy Commission on Docket No. 93-ER-94, August 12, 1994, p. 2.

have made, but were prevented from making because of the damage that was inflicted upon them. These costs relate to courses of action that are not taken and thus are experienced only subjectively. By definition they are unmeasurable.[39]

When the U.S. General Accounting Office looked at this issue for the House Committee on Science, they reached this rather practical summary of the effects to date of forcing use of costs of "externalities" in utility planning:

> The consideration of externalities in the planning process for electricity has generally had no effect on the selection or acquisition of renewable energy sources ... [because] ... electricity from renewable energy usually costs so much more than electricity from fossil fuels that externality considerations do not overcome the difference.[40]

Nonetheless, ludicrous as they seem, calculations of the costs of externalities on the order of magnitude cited by Massachusetts or California are being taken seriously as standards for integrated resource planning. State-approved planning programs are being called "efficient" if they meet this standard, even though that "cost" is vastly higher than the existing retail energy cost or price.

Integrated resource planning also is used as a tool to force the use of alternative "environmentally friendly" energy sources through the use of government forecasts of conventional energy prices that are much higher than eventual actual market prices. The inflated price forecasts distort the comparative valuation of gains from alternative generation and demand scenarios because when federal agencies analyze energy savings potential, they use estimates of future energy costs based on the government's own price forecasts. Indeed, in evaluating such "savings" from government programs,

---

[39]Roy E. Cordato, "Green Pricing of Electricity: The Chimera of Efficiency, the Reality of Politics," unpublished paper, Campbell University, N.C., p. 11.

[40]General Accounting Office, "Electricity Supply, Consideration of Environmental Costs in Selecting Fuel Resources," Washington, May 19, 1995, p. 2.

55

the government may actually be using projected prices even higher than the official (and, thus, already high) forecasts.[41]

The result of the use of such forecasts in planning is to force purchase of alternative sources of generation or to cause other nonenergy investments at higher costs than markets would justify. For example, the California Public Utilities Commission has used $118 per barrel of oil when calculating avoided marginal costs in comparisons between conventionally generated electricity and electricity generated through wind and solar technologies, whereas the current market price for oil is in the $18 to $22 per barrel range.[42] Southern California Edison Company has computed that purchases of "alternative" energy at above market rates cost their customers $2 billion between 1985 and 1993, and would cost them an additional $5.9 billion in the future. High forecasts induced the Tennessee Valley Authority to overinvest by $14 billion in generating facilities. The Minerals Management Service of the Department of the Interior turned down a bid of $26 million for an oil lease based on high government forecasts of future energy prices, then subsequently was able to earn only $15 million for the same lease as prices continued to fall.[43]

Ideologically mandated rising bubble price forecasts not only misstate the future, they lead directly to misspent resources in the present. Embodiment of failed forecasts through requirement of their use for planning, by law, only compounds the problem.

Conclusion: The Bankruptcy of the "Market Failure" Critique

Regulation of electricity and natural gas distributors is necessary neither to control monopoly behavior on behalf of consumers, nor

---

[41]Pointed out by Glenn Schleede in "Energy Price Forecasts Are Leading Business Executives, Regulators, and Other Government Officials to Make Uneconomic Decisions," unpublished working paper, February 20, 1995, p. 10. Schleede notes that the U.S. Department of Commerce, National Institute of Science and Technology document used for such planning, entitled "Energy Price Indices and Discount Factors for Life-Cycle Cost Analysis, 1995," is nominally based on U.S. Department of Energy forecasts, but uses even higher forecasted oil price escalations than published by DOE.

[42]Glenn Schleede, "Illustrations of Costs Resulting from High Energy Price Forecasts That Are Borne by Consumers, Taxpayers, and Shareholders," unpublished paper, March 17, 1996, p. 2.

[43]Data collected by Glenn Schleede, as reported in the source cited in note 41.

to help consumers make cost-effective energy demand reduction investments, nor to internalize the cost of environmental damage.

The belief that electricity and natural gas distributors are natural monopolies that must be regulated to prevent inefficiently high prices and reduced output is contradicted by several forms of evidence. First, historically, regulation of utilities was demanded by some firms within the industry (under the guise of monopoly claims) to suppress unpleasant competition. Second, capital costs may be characterized by economies of scale, but they are a small fraction of utility costs relative to labor and taxes. Thus, total utility marginal costs do not decline with size once a utility has more customers than can be served by a single generating plant, and even below that size competitive access among potential suppliers can reduce costs. Third, the major impediment to the introduction of competition into electric and gas utility markets is often thought to be right-of-way acquisition, but most dwellings are *already* connected to six utility grids, each with an existing separate right of way, each of which could provide competition by a nontraditional producer. For example, cable companies could also transmit electricity via their poles or holes, and vice versa. Finally, just the potential entry of new firms can serve consumer interest regardless of whether actual entry occurs.

Kip Viscusi's work suggests that consumers use discount rates in their energy capital versus fuel use decisions that are very similar to discount rates found elsewhere in the economy. Consumers discount the savings from such purchases at a rate of from 11 to 17 percent because energy prices mostly go down rather than up and thus energy-saving capital investments are risky.

Electric power plants do emit the byproducts of combustion into the atmosphere. In a well-working market the plants should pay for that right, but the alleged cost of externalities created by power plants used in planning proposals under EPACT is too large by a factor of 100. In addition, those externality estimates do not take into account existing pollution reduction efforts in which utilities engage because of concern over tort liability as well as existing environmental regulation.

4. The Pretense of Knowledge: The Effects of Energy Market Regulation

No matter what the intention of the legislature, even seemingly simple and straightforward interventions in markets have ramifications (usually very negative ramifications) far beyond those intended or often even envisioned. The Nobel-laureate economist F. A. Hayek was perhaps the most eloquent and well-known proponent of this argument.[1] Hayek, who began writing about liberty at the time of Hitler and Stalin, was well acquainted with the "fatal conceit" of government planners; that they could intelligently order the economic actions of millions of people without drastic negative consequences. He apparently did not then imagine the depth to which that same conceit had become rooted in American soil.[2]

Indeed, if there is one thing on which most energy scholars now clearly agree, it is that the history of federal and state intervention in energy markets has on the whole proven disastrous.[3] We will not review the issues raised by these authors, as to do justice to the topic would require more space than we have at our disposal in this text. Instead, we will examine perhaps the greatest conceit of

[1]With appreciation and apology to F. A. Hayek, the chapter title is adapted from his 1974 Noble Prize lecture, "The Pretense of Knowledge," reprinted as Chapter 2 of *New Studies in Philosophy, Politics, Economics and the History of Ideas* (Chicago: University of Chicago Press, 1978).

[2]Foreword to the second ed., p. 5, of his *The Road to Serfdom*, originally published in 1944, that ". . . in America, the kind of people to whom this book was mainly addressed seem to have rejected it out of hand as a malicious and disingenuous attack on their finest ideals. The language used and the emotion shown in some of the more adverse criticism . . . received were indeed rather extraordinary."

[3]See Robert L. Bradley Jr., *Oil, Gas & Government: The U.S. Experience*, Vols. I and II (Lanham, Md.: Rowman & Littlefield, 1996) for a comprehensive, interdisciplinary examination of the history of government intervention in oil and gas markets; Stephen McDonald, *Petroleum Conservation in the U.S.* (Baltimore: Johns Hopkins University Press, 1971) for a discussion of state petroleum.

energy regulation: the idea that monopoly regulators can keep prices below those that might otherwise be charged by monopolistic industries. The evidence, in fact, suggests quite the opposite: regulation of monopolies results in higher energy prices than might otherwise be charged by an unregulated monopolist whose territory was not legally protected against entry.

The Averch-Johnson Effect

The idea that a regulated monopoly might spend more than it "needs" to in an economic sense has a name in the economic literature. It is called the "Averch-Johnson effect," from the economists who first studied the possibility.[4] Typically, the return of a regulated company is claimed to be set at a rate comparable to the return that a similar unregulated company in a competitive market also might be able to earn. Superficially, the regulation appears to limit the price below that which an unregulated monopolist might charge, thus setting price at a more competitive level. Professors Averch and Johnson showed that instead, when faced with such regulation, the monopolist has an incentive to invest more in his facility than is needed. This in turn raises cost and thus also price above the competitive level.

Under traditional utility regulation, the amount of profit allowed to the utility is calculated as a percentage of its (public utility commission approved) expenditures. In most companies, cutting costs improves one's competitive posture. In the electric power industry, cutting costs will have the perverse effect of reducing profits.

The A-J effect implies that, if monopoly rents exist in some part of the energy distribution chain, traditional regulation will probably not be able to remove them. In fact, regulation probably is rarely even able to detect them. Because the entire energy utility industry is regulated, there is little evidence to compare the internal practices of regulated companies with nonregulated ones and reveal the excessively costly practices.[5]

---

[4]Harvey Averch and Leland H. Johnson, "Behavior of the Firm Under Regulatory Constraint," *American Economic Review*, vol. 52, pp. 1052–69, December 1962; and Stanislaw H. Wellisz, "Regulation of Natural Gas Pipeline Companies: An Economic Analysis," *Journal of Political Economy*, vol. 71, pp. 30–43, February 1963.

[5]See the classic study of George J. Stigler and Claire Friedland, "What Can Regulators Regulate? The Case of Electricity," *Journal of Law and Economics*, vol. 5, October 1962, reprinted as chapter 14, pp. 224–42, in *The Essence of Stigler*, ed. K. R. Leube and T. G. Moore (Stanford, Calif.: Hoover Institution Press, 1986).

The evidence that does exist suggests that excessively costly practices are maintained by traditional public utility regulation. Consider, for example, a 1986 study that found that the retail price of electricity is *lower* in competitive service areas than in areas "protected" by traditional regulation.[6] Of course, the few utilities that do face competition are still under the control of public utility regulators and thus are not the "unrestrained" monopolists necessary to confirm the A-J effect. Yet the argument that the economies of scale in the industry are such that one company can provide electricity at lower cost than can two or more firms in competition with one another is clearly undercut by such observations.

Further empirical proof that competition would result in lower energy prices than does monopoly regulation was recently provided by the California Public Utilities Commission. In April 1994, the California Commission issued a proposed rule that much of the utility industry described as "the California earthquake."[7] The rule would introduce a limited amount of competition into the retail electric utility markets of California. Part of the reason given by the Commission for this action is their observation of the results of a recent request for bids for electricity. According to the Commission:

> First, the supply of competitively priced power is considerable. The offers for capacity were *six times greater than the amount requested*. Second, the government-sponsored process used to predict "what the utility would do" has little or no bearing on what the market has, and is willing, to offer. Most notable among the results is the remarkable disparity between the price the current planning process estimated customers would have to pay for renewable resources and the actual price offered by the market. In some cases, the market offered renewable electric services for less than one-half of the price estimated by the state. Similar disparities exist between the price the state forecasted for fossil-fired resources and the price the market actually delivered. (emphasis in original)

---

[6]See Walter J. Primeaux Jr., "Competition Between Electric Utilities," in Moorhouse, p. 405. The study matched service territory data on 23 American cities with long histories of competitive provision of electricity, to a similar number of similarly sized and located cities with regulated monopoly services.

[7]"Order Instituting Rulemaking and Order Instituting Investigation," in California Public Utility Commission dockets R.94-04-031 and I.94-04-032, April 20, 1994.

Based on the evidence presented in Chapter 2 about electricity markets before they were regulated, the California conclusions are hardly surprising. Regulation restricts supply and raises prices. Removal of regulation should increase supply and lower price. It is only a surprise because we have tolerated a socialist outlook on energy for so long that when even a small firefly light of competition enters, it appears as a major burst of skyrockets.

Price Regulation: Eliminating or Redistributing Rents?

Regulation leads not only to excessive average prices but to changes in relative prices that would not arise through normal market forces. "Regulatory capture" is the term used to describe the state of affairs in which regulatory agencies become the agents of the parties they nominally control. The historical facts about the electric utility industry encouraging the formation of public utility commissions to stifle competition represent an empirically demonstrated version of the capture theory. The capture theory also predicts that once regulation exists, the process will serve the interests of the organized.

Stigler and Friedland in their classic study comparing regulated with unregulated utilities demonstrated that the ratio of residential retail electricity prices to industrial retail electricity prices was 10 to 20 percent greater if utilities were regulated. The result strongly implies that industrial interests were able to secure relatively lower prices through regulation at the expense of residential customers.[8] Of course, at different times, different groups have been able to "capture" the regulator and turn the results to their benefit. There is of course no economic benefit—indeed there is a cost to the overall

---

[8]See Stigler and Friedland, p. 9. The ratio of residential to industrial price in regulated states was 1.616 in 1917 and 2.459 in 1937, compared with 1.445 in 1917 and 2.047 in 1937 in nonregulated states. If the Stigler-Friedland hypothesis is right, then it is also possible that the progress of increased numbers of states with regulation also caused the ratio to increase, since as more commissions bowed to organized groups, the more easily additional states could also do the same, and in increasing degree. This suggestion is consistent both with the observed change in the ratios and with the fact that regulation increased over the period.

economy—to what becomes little more than a series of perpetual cycles of regulatory revenge.[9]

The time when industrial customers were benefited by regulation is mostly long past. Their trade association, the Electricity Consumers Resource Council, is an energetic supporter of regulatory restructuring. Other parties now receive the economic "rents" of regulation. Two in particular are fuel suppliers and (to no surprise) the government itself.[10]

Under almost all state utility regulatory regimes, electric and gas companies have few incentives to bargain with suppliers in general because the costs are passed on to consumers. This is especially true with regard to fuel suppliers. Commissions regulate fuel costs separately from all other costs in a proceeding typically called a fuel-adjustment, energy-adjustment, or purchased-gas-adjustment clause. In such proceedings, the commission treats fuel costs as a matter of proper cost accounting. But if such costs are not compared with total costs and market conditions, supplier price increases may absorb the monopoly benefits denied the electric or gas distributor. Such regulation simply shifts the monopoly rents from the distributor to the unregulated supplier, perhaps as take or pay contracts, automatic price escalator clauses, automatic pay increase clauses (or other supplier employee job protection) unrelated to productivity, or in other forms.[11]

Government itself has a vested interest in the continued existence of energy monopolies because monopolies can pass on to consumers the costs of explicit taxes and social policies ordered by the government. Chapter 2 demonstrated that taxes are a larger fraction of

[9]See George J. Stigler, "The Theory of Economic Regulation," *Bell Journal of Management Science and Economics* 2, Spring 1971, pp. 3–21; William A. Jordan, "Producer Protection, Prior Market Structure and the Effects of Government Regulation," *Journal of Law and Economics* 15, April 1972, pp. 151–76; Richard A. Posner, "Theories of Economic Regulation," *Bell Journal of Management Science and Economics* 5, Autumn 1974, pp. 335–58; and Sam Peltzman, "Toward a More General Theory of Regulation," *Journal of Law and Economics* 19, August 1976, pp. 211–48.

[10]For a similar argument about the effect of oil-price regulation on the petroleum market, see Joseph P. Kalt, *The Economics and Politics of Oil Price Regulation* (Cambridge: MIT Press, 1981).

[11]See Paul A. Ballonoff, "Is There a Role for Regulators in a Deregulated or Less Regulated Economy?" pp. 525–32 in Robert Trappl (ed.), *Cybernetics and Systems '88* (Boston: Kluwer Academic Publishers, 1988).

retail utility prices than capital cost recovery. This is a peculiar result for an industry supposedly regulated to "protect" consumers from excessive prices due to natural monopoly. The most optimistic view of the large percentage of taxes in the average gas or electricity bill is that, if utility services are natural monopolies, then the effect of government regulation has been to transfer the monopoly benefits— the economic rents—to government rather than to limit price to consumers.

Government also absorbs the proceeds of monopoly rather than eliminates them through various expensive planning policies imposed on utilities, such as the requirement for "integrated resource planning" that was imposed by the *Energy Policy Act* of 1992. Under "least cost planning" utilities are forced to impose higher cost energy choices on consumers than consumers voluntarily choose themselves, and then to pass on this excess cost by absorbing it into consumer energy payments.

The Unintended Consequences of Energy Regulation

Regulation not only fails to control "excessive" pricing (the alleged rationale for its existence), but distorts choices as well. This section describes two instances in which energy regulation has induced choices that are inefficient and would not have occurred in the absence of governmental persuasion; a third case in which an efficient choice is discouraged by the existence of traditional utility regulation; and a fourth case that will create distortions in the future if enacted.

The Environmental Costs of Electric Car Mandates

Many environmentalists advocate the use of electric cars rather than gasoline powered vehicles.[12] Perhaps the immediate result of turning on a light switch seems "clean" compared with, say, lighting

[12]In fact, they succeeded in inserting language in the 1990 Clean Air Act amendments that mandates the use of zero emission vehicles (ZEV) in California at the rate of 2 percent of sales in 1998 (about 20,000 cars per year), rising to 5 percent of sales in 2001. Other states can adopt the more stringent California emission standards if they wish, and Massachusetts and New York have done so. In December 1995 the California Air Resources Board drafted proposed changes to the ZEV mandate that would reduce the required sales from 60,000 electric cars to 3,750 over the three years 1998 to 2000. See Lawrence M. Fisher, "California Is Backing Off Mandate for Electric Car," *The New York Times*, December 26, 1995, p. A14.

a kerosene lamp in the living room. But electricity is created by consuming other fuels. Switching from petroleum-powered cars to electric ones simply moves the point of primary fuel consumption from the vehicle to the electricity generation unit. Each kWh of electricity contains 3,413 Btu of usable energy. The most efficient electricity generators use about 8,000 Btu to create 1 kWh of electricity; typical electric utility systems require an average of about 10,000 Btu to create a kWh of electricity.

Thus, the fuel efficiency of electric energy can be no better than about 30 to 40 percent of the energy contained in the initial fuel, before transmission and distribution line losses or the efficiency of the use of energy in the vehicle are even measured. Such energy efficiency losses might be justified if the environmental benefits in air quality were large. But the electric plant may not be "cleaner" in the environmental sense: the pollution produced for an equivalent amount of energy by electric generation from a coal plant can be three times worse than the emissions of a gasoline-powered car, or equal to the emissions of a natural gas-powered vehicle.[13] Electric cars do not provide environmental benefits to offset their energy losses. In fact, in the short run increased use of electric cars could drastically increase the emissions of lead in the environment because of the use of conventional lead-acid batteries.[14]

The Energy Policy Act of 1992: Sabotaging Natural Gas Vehicles

In title IV of the Energy Policy Act of 1992, Congress limits regulation of natural gas when it is sold as a vehicle fuel. But this cannot realistically be achieved while traditional utility regulation of natural gas remains in place. Even if delivered vehicular natural gas price were unregulated, many or all of the components of that cost are effectively regulated if the gas is delivered by an otherwise regulated utility or its affiliate.

This "effective regulation" occurs for two reasons. First, a regulatory commission will still be able to allocate costs of the utility

---

[13]See Kenneth T. Derr, "Alternative Vehicle Fuels Do Not Offer Viable Alternative to Gasoline in the United States," *Oil and Gas Journal*, December 19, 1994, p. 32.

[14]See Peter Passell, "Lead-Based Battery Used in Electric Cars May Pose Hazard," *The New York Times*, April 9, 1995, p. A1; and Peter Passell, "Economic Scene," *The New York Times*, August 29, 1996, p. C2.

company between regulated and unregulated services. A commission will typically do this in the normal course of its rate review of the regulated utility. It will be difficult to prevent regulators from allocating these costs in a manner that subsidizes regulated activities. After all, the regulators are also politicians, and politicians receive rewards by benefiting constituents (those who receive regulated or jurisdictional services), not by benefiting others (those who receive nonregulated or nonjurisdictional services).

Second, if prices for natural gas used in vehicular fuel differ from prices for regulated uses, regulators will inquire about the nature of any cost differences that justify a divergence between the regulated and unregulated services. In such circumstances, and in the heated public atmosphere of such proceedings, it will be very difficult for the affected utilities to explain what might make regulated natural gas cost higher than unregulated supply.

If the regulated company cannot satisfy the regulator about such differences, the most likely result is that the company will have to absorb them. The prudent natural gas distributor will purchase gas, for both regulated and unregulated services, in a manner likely to satisfy the regulator. Therefore, as a practical matter, the "unregulated" natural gas vehicle fuel business will also be largely regulated in the same manner as is the (officially) regulated utility business. Retail delivery of "unregulated" natural gas used as a vehicular fuel while other retail natural gas remains regulated thus actually may inhibit, and certainly will not promote, use of vehicular natural gas. It even may have the still more paradoxical result of discouraging the local natural gas utility from entering the vehicle fuel business simply because the market characteristics of vehicle fuels and retail distribution for home and commercial use are different in ways that energy regulators seem genetically to misunderstand.

Energy Taxes: A Monopolist's Best Friend

Traditional regulatory myths are pernicious in the arena of energy tax policy. The myth that energy distribution is a natural monopoly frustrates an analysis of how taxes affect the consumption of energy. Certainly, any consumption tax has the effect of "conserving" resources, because the tax becomes a cost that is passed along to consumers, and higher prices lead to a decline in the amount purchased. But the ability of energy companies to pass taxes on to consumers is overestimated because of the incorrect belief that energy companies are magical monopolies that can pass on to customers the total effect of taxes on their product. Instead, taxes on energy consumption, like taxes on the consumption of any product, are shared by consumers and producers in proportion to the sensitivity (elasticities) of supply and demand to changes in prices.

Any energy consumption tax reduces profits to producers to some degree regardless of the point at which the tax is applied. Thus, even if only a tax on energy consumption were imposed, some part of that tax would be transferred to producers. Given some reasonable assumptions about the shapes of petroleum industry supply and demand curves, perhaps about one-third of a consumption tax will be passed back in the form of reduced price to producers, because the total, delivered, consumer point-of-consumption price likely will increase by only two-thirds of the taxed amount.[15] The most recently proposed American Btu tax was about 26 cents per MMBtu. Assuming the one-third, two-thirds division, approximately 9 cents of a 26-cent per MMBtu tax on natural gas would be paid by producers; if the ratio were half and half, for example, then producers would absorb even more, about 13 cents from a 26-cent tax.

This 9 cents would likely affect production decisions. The current average gross margin on production is about 45 cents. That average, however, is the average of costs among very diverse companies. Some producers have finding costs that are five times higher than those of the low-cost producers, and their development costs are also significantly higher. For some producers their entire gross margin may be only 9 cents. For all producers, certainly, 9 cents is a significant part of gross margin. In a $2 market, 9 cents represents almost 5 percent of total revenue. Compare this to the portion of total revenue for net income of all petroleum producers. In 1991, according to data of the Energy Information Administration, the total net income of all sampled companies was $14.7 billion of a total revenue of $469.3 billion, or about 3 percent.[16]

---

[15]The one-third to producer, two-thirds to consumer allocation is simply an illustration. If a larger portion of the tax is passed to producers, the effects illustrated here are more profound.

[16]U.S. Department of Energy, Energy Information Administration, *Performance Profiles of Major Energy Producers, 1991*, December 1992.

The effect of an energy consumption tax on production will therefore be perverse. Because producers make decisions on new production based on expected future profits from that new production, a 9-cent tax, higher than the income of many producers, will cause some producers to reduce production. As low success (that is, high cost and, hence, low return) suppliers drop out of the market, they may sell their reserves to producers with higher success rates. That may lower the overall level of supply costs. But as it occurs, market concentration will increase. Because of normal market operations, governments cannot impose taxes that increase prices to consumers without affecting economic returns and market structure along the entire chain of production, distribution, and subsequent use. A Btu tax could induce the very monopolistic producer industry structure domestically that the tax is purportedly intended to remove internationally.

It is also true that some part of a tax at any part of the distribution chain is necessarily passed on to consumers. Therefore, reducing the present tax burden on energy—say only to that placed on the rest of the economy—will cause retail prices to fall and the prices paid to producers to rise. Thus, if encouraging domestic energy production and reduction of foreign imports were government policy goals, one means to reach it would be to reduce energy taxes, not to increase them.

Regulating Holding Companies: The Cost to Consumers

Now let us consider how traditional public utility regulation, as embodied in the Public Utility Holding Company Act of 1935, inhibits innovations in the organization of gas and electric utilities that would reduce energy costs for consumers. Title I of the 1935 act[17] (often shortened to PUHCA) was intended to control the existence and reach of corporations that were used to control public utility operating companies. The operating companies were generally local in nature, and thus within the reach of the local state commissions for retail transactions or the federal commissions for interstate transactions. But the ownership structures of the operating companies

---

[17]49 Stat. 803 (1935), or 15 U.S.C. 79a and following.

were seen as beyond state control, especially where a holding company owned operating companies in disparate regions of the country.

The Act prohibited a single holding company from controlling firms unless they were operated as single coordinated and integrated systems within a single state. Regulated companies also could not offer services that would compete directly with other local utilities. Electric companies could not compete with local natural gas companies by offering natural gas service, for example.

The restrictions of PUHCA result in organizations that are almost certainly not efficient. Richard Geddes says:

> granting exclusive monopoly territories does not assure that firms can operate at their optimal size. Firms might grow larger under a less restrictive regulatory framework and thus reap greater benefits from scale factors in both coordination of power production and transmission.[18]

PUHCA prevents this. Relatedly, Vernon Smith notes:

> early in the century when the industry was in its infancy, most electric power was generated locally or transmitted over relatively short distances (100 miles or less). The exclusive monopoly territories granted in state certificates reflected the service areas of these local, self-contained power systems.[19]

If service territories today still largely reflect these obsolete market conditions, state-assured territories are not efficiently sized.

Conclusion: The "Fatal Conceit" of Regulation

The actual effects of regulation usually are not the same as the intentions of the legislature that enacted the statute. The equity

[18]Richard Geddes, "A Historical Perspective on Electric Utility Regulation," *Regulation*, vol. 15, no. 1, Winter 1992, p. 81. See also his "Time to Repeal the Public Utility Holding Company Act," *Cato Journal*, vol. 16, no. 1, Spring/Summer 1996. For similar analyses of PUHCA, see Richard L. Gordon, "Deregulation and the New Competitive Order in Electricity Generation: The Stumble Towards (Possible) Free-Market Generation," unpublished paper, 1995; Paul L. Joskow, "Expanding Competitive Opportunities in Electricity Generation," *Regulation*, vol. 15, no. 1, Winter 1992, pp. 25–37; and Richard L. Gordon, "The Public Utility Holding Company Act, The Easy Step in Electric Utility Regulatory Reform," *Regulation*, vol. 15, no. 1, Winter 1992, pp. 58–65.

[19]Vernon L. Smith, "Currents of Competition in Electricity Markets," *Regulation*, vol. 10, no. 2, 1987, p. 24.

improvements that often motivate the legislature to act do not materialize. Worse, the government encourages choices that hurt consumers and discourages choices that would aid them. In addition, even if regulation actually enhances market choices at one time, regulation often later inhibits useful innovations because markets evolve more quickly than the bureaucratic apparatus created to regulate them. Thus Pasour tells us:

> ... abolishing the statutory monopoly enjoyed by public utilities to permit freedom of entry may be a more effective means of increasing competition than attempts to modify or "fine-tune" current methods of regulation. ... Government intervention to regulate price should be evaluated in terms of a principled approach, rather than a case-by-case approach.[20]

And, citing Yeager,

> If we avoid appraising and comparing economic systems as wholes, if we avoid forming and acting on a coherent conception of a good society, we shall make momentous choices by ignorance and default. The opposite approach, respecting principles, would go far ... toward reinstating the wisdom of the Founding Fathers regarding the scope and power of government.[21]

The problem is not that limited government could regulate better if it were instead unlimited—that solution is known from experience, often bitter experience, to fail. The problem is the opposite: even limited intervention by government cannot easily, indeed, most often not at all, be directed toward intended results. After a half century of centralized monopoly control of energy utilities, Britain has recently begun to acknowledge the mistake and undo it:

> Regulating a market which is actually or potentially competitive is not an additional safeguard for consumers but a recipe for stifling initiative and hampering competition. ... The fundamental problem is that, if there is no competitive market,

---

[20]E. C. Pasour, "Information: A Neglected Aspect of the Theory of Price Regulation," *Cato Journal*, vol. 3, no. 3, Winter 1983–84, pp. 866–67.

[21]Leland B. Yeager, "Economics and Principles," *Southern Economic Journal*, vol. 42, April 1976, pp. 569–70, quoted in ibid., p. 867.

regulators cannot know what to do; they have no standard on which to base their actions. But if there is a competitive market, regulation becomes redundant.[22]

Government is far better at preserving wrong answers than fostering new ones. Rather than further expanding, we need to better respect and acknowledge the real limitations of government.

[22]Pages 16–17 in Colin Robinson, "Gas: What to Do after the MMC Verdict," pp. 1–19 in M. E. Beasley, ed., *Regulating Utilities: The Way Forward* (London: London Business School Institute of Economic Affairs, 1994). The conclusion is also cited in Israel Kirzner, "The Perils of Regulation: A Market Process Approach," in I. Kirzner, *Discovery and the Capitalist Process* (Chicago: University of Chicago Press, 1985).

5. The Constitution in Exile?

A rigorous examination of the United States Constitution finds no constitutional legitimacy for federal—or even most state—energy regulation. This chapter examines various constitutional rationales that have been, or might be, offered to justify energy regulation. The clause in the preamble of the Constitution stating that one of the purposes of the Constitution is "to promote the general welfare" is often used to justify federal regulation of the economy. Yet a close reading of the Federalist papers as well as Madison's notes about the Constitutional Convention does not support the view that the "general welfare clause" is an open-ended blank check for federal intervention.

In fact, the Constitution was written with the market revolution in England in mind, in which the power of the Crown to establish economic monopolies was severely restricted by the Statute of Monopolies. States, unlike the federal government, may intervene in their economies but only so long as the intervention does not restrict interstate competition. These three lines of reasoning suggest that current regulation of utilities by commission is unconstitutional.

The General Welfare Clause: A Blank Check?

One of the more pervasive sources of present federal regulation is the claim that the enumerated powers of Congress under the commerce clause enable the Congress to legislate essentially anything in the name of the general welfare. The commerce clause at Article I, section 8, paragraph 3 reads: "The congress shall have the power to regulate commerce with foreign nations, and among the several States, and with the Indian tribes." That enumerated power allegedly allows federal regulation of markets because the Preamble to the Constitution reads, in part, "We the people of the United States, in order to . . . promote the general welfare . . . do ordain and establish this Constitution," and because Article I, section 8 has similar language also in the initial paragraph of that section.

But the so-called general welfare clause of the Constitution is not a blank check that empowers the federal government to do anything it deems good. It is instead a general introduction explaining the exercise of the enumerated powers of Congress that are set forth in Article I, section 8. Supreme Court Justice Joseph Story, who sat on the Court from 1811 until 1845, was an ardent nationalist who promoted constitutional interpretations that expanded federal power. In a work explaining the Constitution to the general public,[1] Justice Story devotes his entire discussion of the so-called welfare clause to how the commerce power is necessary to prevent states from interfering with interstate and international commerce through the adoption of nonuniform import duties. Justice Story's commentary does not suggest that the welfare language is anything other than a mere explanatory preface to the necessity of federal power to limit state interference with commerce.

If a more general federal power had been intended, surely Justices like Story, who were seeking to justify central power, would have made such arguments. They would cite them because a general welfare power, if it existed, would dwarf the highly limited regulatory authority that flows from the commerce clause. Indeed, if the general welfare clause really was pregnant with such power, why then bother with a list of specifically enumerated powers at all? A page or two would have sufficed, for all the Constitution would have needed were those parts that established the three separate branches of government and the general welfare clause. Article I, section 8—which enumerates the powers of the legislature—would be superfluous. The conventional wisdom in constitutional law that the federal government has some allocated power to "promote the general welfare" in the broadest sense of those words is an incorrect interpretation.

Further support for this view comes from the Federalist papers, brief contemporary essays written to support ratification of the Constitution by no less than John Jay, Alexander Hamilton, and James Madison. The most relevant are Papers 41 through 44 because of their content—analysis of the commerce clause including the welfare

---

[1]See pp. 64–65 in Joseph Story, *A Familiar Exposition of the Constitution of the United States* (New York: Harper & Brothers, originally published 1859, reprinted 1986, Regnery Gateway).

clause language—and their authorship, James Madison. Madison is well known as an author of the Constitution as well as the primary source of concurrent notes of the debates of the Federal Convention (today called the "Constitutional Convention") of 1787.[2] The various sections of proposed language can be found in the Madison notes, especially from August 18, 1787, in which he documents what was submitted to the Convention for consideration related to various enumerated specific powers of Congress. As is discussed further hereafter, many possible powers were enumerated in proposed texts, and not adopted. Madison's views on these alternatives are contained in Federalist papers numbers 41 through 44, and are entirely consistent with those of Justice Story.

In Federalist No. 41, Madison summarizes the relationship of the general preface language, including the "welfare" language, to the subsequent more detailed enumeration of specific powers, as follows:

> Some who have denied the necessity of the power of taxation [to the federal government] have grounded a very fierce attack against the Constitution, on the language on which it is defined. It has been urged and echoed that the power to "lay and collect taxes, duties, imposts, and excises, to pay the debts, and provide for the common defense and general welfare of the United States" amounts to an unlimited commission to exercise every power which may be alleged to be necessary for the common defense or general welfare. No stronger proof could be given of the distress under which these writers labor for objections, than *their stooping to such a misconstruction.* (emphasis added)

Thus Madison, who like Story after him sought to defend federal power, treats with derision the claim of opponents of federal powers that the welfare clause is a general grant of power. Madison continues Federalist No. 41 in this language of angry paradox:

---

[2]James Madison, *Notes on the Debates of the Federal Convention of 1787*, reprint edition (New York: W.W. Norton, 1987). A thoroughly documented discussion of Madison's views of the welfare language is given by Leonard R. Sorensen, *Madison on the "General Welfare" of America, His Consistent Constitutional Vision* (Lanham, Md.: Rowman & Littlefield, 1995). My discussion differs from Sorensen's because I wish not only to show the limited nature of the welfare clause, but also to compare the "welfare," "commerce," and "patent" clauses.

For what purpose could the enumeration of particular powers be inserted, if these and all others were meant to be
included in the preceding general power? Nothing is more
natural or more common than first to use a general phrase,
and then to explain and qualify by an enumeration of the
particulars. But the idea of an enumeration of particulars
which neither explain nor qualify the general meaning, and
can have no other effect than to confound and mislead, is
an absurdity. What would have been thought of that assem-
bly, if, attaching themselves to these general expressions and
*disregarding the specifications which limit their import, they had
exercised an unlimited power of providing for the general welfare?*
(emphasis added)

Earlier in Federalist No. 41, Madison grouped the powers granted
the federal government into six "classes," which were "1. Security
against foreign danger; 2. Regulation of the intercourse with foreign
nations; 3. Maintenance of the harmony and proper intercourse
among the States; 4. Certain miscellaneous objects of general utility;
5. Restraint of the states from certain injurious acts; 6. Provisions for
giving due efficacy to all of these powers." Note that there is no
listing of "general welfare" as one of these general classes.

In Federalist No. 42 especially, Madison discusses the third and
fifth classes, which relate to the states. As to the third, "Under this
head might be included the particular restraints imposed on the
authority of the States . . ." so that number 3, which sounds like a
grant to intervene, is actually a listing of things the states cannot
do, coupled with a very specific list of things the federal government
is specifically authorized to do, such as to establish posts and roads,
or a uniform system of weights and measures. Promoting welfare
is not in Madison's list of things that promote harmony among the
states in Federalist No. 42.

As to "the necessity of superintending authority over the recipro-
cal trade of confederated States," he gives as examples the problems
of local taxation within the Cantons of Switzerland, and similar
constraints imposed against this in Germany and The Netherlands.
So, consistent with the explanations of Justice Story, Madison saw
the commerce clause as a restraint on the powers of the states. Thus
Federalist No. 42 follows the theme of derision of No. 41, showing
more rationally that an enumerated list of constraints on the states

is certainly not anything like a general grant to do anything in the name of the general welfare.

Madison discusses his fifth class of federal powers, "restraint of the states from certain injurious acts," in Federalist No. 41. Again, Madison makes no statement remotely implying the federal government has some general power over the general welfare. Instead, he continues his theme of enumerating ways in which the Constitution constrains states. For example, regarding the constitutional restraint on states from laying duties for imports or exports except "what may be absolutely necessary for executing its inspection laws," he explains:

> The manner in which the restraint is qualified seems well calculated at once to secure to the States a reasonable discretion in providing for the convenience of their imports and exports, *and to the United States a reasonable check against the abuse of this discretion.* (emphasis added)

In Federalist No. 43 Madison discusses other powers including the "power to promote the progress of science and the useful arts by securing for a limited time to authors and inventors the exclusive right to their respective writings and discoveries," without any use of the word "patent," although he does allow himself use of the words "copyright of authors." The importance of the omission of the word "patent" from the Constitution is the subject of the next section.

It is abundantly clear that the Constitution does not contain a grant of power to the federal government in the form of a "welfare clause," and also that the commerce clause is not, by any stretch of the imagination, a general grant to regulate all commerce, and especially not a grant for the federal Congress to legislate generally in the name of the general welfare.

Allocating Economic Privileges: The Dubious "General Patent Power"

The regulatory regime currently governing energy utility and business regulation, more generally, is implicitly premised upon the exercise of a "general patent power," the power to allocate economic privileges. An examination of its historic origins and legal lineage, however, makes clear that the general patent power is one power that the Founding Fathers expressly prohibited the federal government from ever exercising. While this observation does not necessarily prevent economic regulation per se, it does imply that the means by which utility commissions regulate today, especially federal regulatory commissions, is blatantly unconstitutional, and the actions of state commissions should be greatly limited.

Today, the term "patent" refers to the right given to inventors and authors to sell their ideas or creations exclusively for a limited time. Although the word "patent" does not appear in the Constitution, the ability of Congress to grant such exclusive rights to authors or inventors is one of the powers of Congress specifically enumerated in Article I, section 8, "The Congress shall have the power ... to promote the progress of science and the useful arts, by securing, for limited times, to authors and inventors the exclusive rights to their respective writings and discoveries."

When the Constitution was written, however, the word "patent" had a much different and broader meaning. Justice Story, for example, provides a definition of the term in a glossary included in his popular exposition of the Constitution:

> *Patent*, an abbreviated expression, signifying letters patent, or open letters, or grants of the government, under the seal thereof, granting some right, privilege, or property, to a person who is thence called the Patentee. Thus the government grants the public lands, by a patent, to the purchaser. So, a copy-right in a book, or an exclusive right to an invention is granted by a patent. When the word patent is used in conversation, it ordinarily is limited to a patent-right for an invention.[3]

Story was simply reflecting a centuries-old accepted use of the term in English law. William Blackstone, who wrote the classic and still-cited work on English law of the 18th century, tells us that

> The King's grants are a matter of public record.... These grants, whether of lands, honors, liberties, franchises, or ought besides, are contained in charters, or letters *patent*, that is, open letters, *literae patentes*: so called because they are not sealed up, but exposed to view....[4]

---

[3]Story.

[4]William Blackstone, *Commentaries on the Laws of England*, Vol. II, 1766, p. 346, reprint edition (Chicago: University of Chicago Press, 1977).

The term general patent power, as used in this book, therefore, refers to the meaning of the term "patent" as used in the 16th to 18th centuries, a grant of any economic right by government.

Now, what is particularly interesting about this more general definition is that no "patent clause" exists in this sense in the Constitution. There is no general explanation of federal powers using either the term or the general expression of such power. Even in Article I, section 8, paragraph 8, the term "patent" is not used when describing the powers of the federal government to grant exclusive rights to authors or inventors. Instead, the explicit definition of the specific power to grant rights to authors or inventors is used. Since Article I, section 8 listing the powers of Congress is simply a specific enumeration of powers, since the only constitutional text that uses any part of the possible powers described by "patent" enumerates only a specific form of that power, and since Madison's notes on the Constitutional Convention as cited earlier show that broader terminologies were considered and rejected, we must logically conclude that the Constitution intended that the federal government would have no such general patent power.

This fact is also abundantly clear from the contemporary explanations of Madison in the Federalist papers, but to reinforce this fact, shortly after adoption of the Constitution, the Tenth Amendment was added, reading, "The powers not delegated to the United States by the Constitution, nor prohibited by it to the States, are reserved to the States respectively, or to the people." Clearly, the advocates of the Constitution envisioned the Tenth Amendment as an exclamation mark for the limitations implied by the structure of Article I, section 8, stipulating, in effect, "and we really mean it."

The Crown's "Government of Patentees"

It is not surprising that people who intellectually descended from the English Opposition would write a constitution that specifically denies the general patent power, prevents the government from allocating economic rights, and grants only specific limited-term patents to encourage science and useful arts. That very policy had been law in England for 150 years at the time of the American Revolution, and indeed had been adopted by the colonies in their own laws. The authors of the Constitution simply embodied the accepted English law.

The English Parliament's pro-market revolution occurred in 1624. King James I, following in his predecessors' steps, had made a habit of granting exclusive rights for all manner of economic activity. Typically rights, called "patents," were for patents of monopoly on economic activity, but they were also sold as exclusive rights for the exercise of government offices or other activities. Patents often were sold for their revenue benefit to the Crown, whatever their commercial merits as monopolized services. The sale of government offices was so widespread and abused that the practice was criticized at the time as being "government by patentees."[5]

Perhaps the sale of the rights was good business for the Crown, but eventually the Parliament no longer accepted rampant monopolization created by what today we would call "the administration." The period 1621 through 1624 was dominated by issues of how to control Crown powers, especially those in finance, and thus focused heavily on the issue of sales and use of patents of monopoly. In 1624 Parliament passed one of the first statutes specifically limiting the prerogatives of the Crown.[6] Parliament's anti-monopoly law was in many ways like the modern antitrust laws, but with a critical difference. Modern American antitrust laws are named for one of their purposes: to limit private trusts. The Parliament's Statute of Monopolies of 1624 was instead specifically directed at limiting government grants.

Consider these words from that Statute:

> ... upon misinformations and untrue pretenses of the public good many such grants ... [of patents of monopoly from the Crown] have been unduly obtained and unlawfully put into execution, to the great grievance and inconvenience of your Majesty's subjects and contrary to the laws of your realm. To prevent the like in time to come, be it enacted that all monopolies heretofore or hereafter to be granted to any person or persons, bodies politic or corporate, for the sole buying, selling, making, working or using any thing within

---

[5] See for example Barry Coward, *The Stuart Age* (London: Longman, 1980), especially Chapter 5: "The Breakdown of the English Constitution, 1621–1640," p. 133; or John Brewer, *The Sinews of Power* (Cambridge: Harvard University Press), 1990, p. 17.

[6] See J. P. Cooper, "The Fall of the Stuart Monarchy," pp. 531–84 in *The New Cambridge Modern History, Volume IV* (Cambridge: Cambridge University Press, 1970), especially at pp. 546–53.

this realm, or any other monopolies, are and shall be utterly void and of no effect and in no wise to be put into execution.[7]

Thus, the Parliament created a general policy that the government may not grant exclusive rights, known as "patents" according to the usage of the term at the time. The law then made an exception for what today we refer to as "patents":

... provided never the less that the above [prohibition] shall not [be] extended to any letters patents and grants of privilege for the term of 21 years or under, ... to the first true inventor of such manufacture.[8]

The necessity of such a statute is illustrated by an exchange that occurred in the English Parliament in 1601 between Member Sir Robert Roth and others named and unnamed:

There have been divers patents granted since the last Parliament. There now in being viz. the patents for currants, iron, powder, cards, horns, ox shin bones, train oil, lists of cloth, ashes, bottles, glasses, bags, shreds of gloves, aniseed, vinegar, sea-coals, steel, aquavitae, brushes, pots, salt, saltpetre, lead, accedence, oil, transportation of leather, calamine stone, oil of blubber, fumothoes, or dried pilchard of smoke, and divers others.

Upon reading of the patents aforesaid Mr. Hackwell of Lincoln's Inn stood up and asked this: "Is bread not there?"

"Bread?" quoth another.

"This voice seems strange," quoth a third.

"No," quoth Mr. Hackwell, "but if order be not taken for these, bread will be there before the next Parliament."[9]

The reference to bread is ironic, but the issue was not, as the Statute of Monopolies proved in 1624. Clearly, the proliferation of government control of commerce through uncontrolled grants of patents by the Crown was seen as an affront to the liberty of the people.

---

[7]21 and 22 Jac. Ic. An Act Concerning Monopolies and Dispensations with Penal Laws and the forfeiture thereof, also known in other literature and cited in this text as the Statute of Monopolies.

[8]See previous citation, paragraph V of the Statute.

[9]Joel Hurstfield and Alan G. R. Smith, *Elizabethan People* (London: Unwin Brothers, 1972), p. 62, reprinted there from the *Historical Collection* of Heywood Townshend, 1680.

Why do we care about this ancient statute? The Statute of Monopolies was a critical document in the evolution of the American theory of limited government. The charters of the colony companies, themselves, were often general patents. But by 1641 the Massachusetts General Court had begun the practice, following the Statute of Monopolies, of granting only limited, term-specific patents of monopoly and only for new economic development.[10]

These lessons were certainly not forgotten at the time of the American Revolution. Shortly after our own Constitution was written, Justice Story tells us that a patent is a "grant of the government ... granting some right ... to a person ...," which is to say, a government grant of an exclusive right. In particular, in the definition given by Justice Story, the term refers to an exclusive economic right, in a way that includes the general patents specifically prohibited to the Crown by the English Parliament. The "patent clause" of the U.S. Constitution confers only a limited ability to create exclusive rights, restricted to grants to writers and inventors for limited times, the same exception to denial of general patent power made in 1624 by Parliament.

The "patent clause" of the U.S. Constitution also does not use the word "patent" at all, a significant and intentional omission. People

[10]One of the most detailed histories of the Statute of Monopolies can be found in a book not on English history, but on New England colonial history. See especially Chapter 5, "Civil Society and Economic Development," in Stephen Innes' *Creating the Commonwealth, The Economic Culture of Puritan New England* (New York: W.W. Norton and Company, 1995). Until recently, the issues addressed by the Statute of Monopolies were rather familiar to European analysts of western society until the early 20 century. Even Max Weber, the famous sociologist, notes them (see p. 284 in Max Weber, *General Economic History* [New Brunswick, N.J.: Transaction Publishers, 1981]):

> In the 16th and 17th centuries an additional force working for the rationalization of the financial operations of rulers appeared in the monopoly powers of princes. In part they assumed monopoly powers themselves and in part they granted monopolistic concessions, involving of course the payment of notable sums to the political authority.... The policy was most extensively employed in England and was developed in an especially systematic manner by the Stuarts, and there it also first broke down, under the protests of Parliament. Each new industry and establishment of the Stuart period was for this purpose bound up with a royal concession and granted a monopoly.... But these industrial monopolies established for fiscal purposes broke down almost without exception after the triumph of Parliament.

of the time understood the word as having a broader meaning than given today. Therefore, by specifically failing to use that word they also intentionally and specifically denied such broad power to the newly created United States's national and federal government.[11]

Consider again the United States Constitution at Article I, section 8, the commerce clause, which says that Congress has the power "to regulate commerce with foreign nations, and among the several States, and with the Indian tribes." Congress can regulate commerce among the several states, but has no general power of the patent except in a very limited area—to grant rights to inventors and authors. Therefore, *Congress also has no power to regulate commerce among the several states by the use of an exclusive grant.* It does not have that power because it was specifically and intentionally omitted from the list of powers of the federal government by, among other devices, not even using the word "patent" in granting the very narrow patent power that the federal government does possess.[12]

The Constitutional Convention and the Rejection of Regulatory Power

The failure of the Constitutional Convention to adopt language authorizing the government to issue general patents, or anything

[11]Contemporary historians of England also recognize the broader meaning of "patent" as equivalent to any government-granted monopoly. For example, in the index of recently published history, Conrad Russell's *The Fall of the British Monarchies 1637–1642* (New York: Oxford University Press, 1991), the word "monopoly" is listed with several page cites, but the word "patent" does not appear at all. In fact, on one of the cited pages for "monopoly," p. 81, the word "monopoly" does not appear; instead that page refers to an action of the king to recall a number of "unpopular patents" as amelioration of other actions desired by the Crown. In the index to the work of Innes on New England history (cited in the previous footnote), this identity of meaning is even more explicit: the citation under "patent" says simply "see monopoly."

[12]The same general argument also explains why the States, not the federal government, charter corporations. The power to create a corporation is a remnant today of the general patent power of the Crown in 1600. In part through the Statute of Monopolies, Parliament eventually assumed this power itself. The power to charter corporations is not among the enumerated powers of the federal Congress, the national legislature, and thus remained with the States. In fact, the power to charter corporations was among the specific powers proposed on August 18, 1787, in the list of proposals considered and rejected, as documented by James Madison in his *Notes*. The specific language considered and rejected is:

> To grant charters of incorporation in cases where the public good may require them, and the authority of a single State may be incompetent.

similar, tells us three important things. First, like the English Parliament of 1624, the Convention rejected the use of claims of the general good as a basis for legislation by Congress. Madison's derision of claims that the welfare language is a general grant of power is also supported by the Convention's rejection of specific language that might have granted such general power.

Second, the Convention specifically considered a form of power that allowed Congress to act when it deemed some matter of public good to exist and when "the authority of a single State may be incompetent." The Convention also rejected that language. That is, the Congress has no power to act merely because some matter of public good is demonstrated in some fashion, and even if it is believed the states have no ability to act on the circumstance. Clearly the authors of the Constitution knew their history, such as the abuses that led to the Statute of Monopolies, and sought to avoid them. That particular deliberate omission suggests that the Convention foresaw, from knowledge of history, the kinds of justifications often used today for creating statutes under the (nonexistent) welfare clause and intentionally excluded them.

Third, the Convention specifically considered other aspects of the general patent power, and rejected them. Before the final "patent" clause was actually adopted, other proposals were considered, but not accepted, by the Constitutional Convention. One text not adopted related to the power to create universities:

> To secure to literary authors their copy rights for a limited time, To establish a University, To encourage, by proper premiums and provisions, the advancement of useful knowledge and discoveries.[13]

Note that this language uses the term "copyrights," but does not use the term "patent." Also, note that this proposed language, by virtue of its rejection, indicates that the Convention clearly not only decided against giving the federal government power to create a national university but also denied the federal government the

---

[13]See I. Bernard Cohen, *Science and the Founding Fathers* (New York: W.W. Norton and Company, 1995), especially pp. 237–43, as well as Madison's *Notes* for August 18, 1787.

power to use "proper premiums and provisions" to encourage science.[14]

Another rejected text also included more expansive powers:

> To establish seminaries for the promotion of literature and the arts and sciences, To grant charters of incorporation, To grant patents for useful inventions, To secure to Authors exclusive rights for a certain time, To establish public institutions, rewards and immunities for the promotion of agriculture, commerce, trades and manufactures.[15]

This version uses the term "patents," but not alone; the term occurs in the phrase "patents for useful inventions," a reference to a similar exception in the Statute of Monopolies. This phrase would have been unnecessary had the term "patent" had only the modern meaning. The text does not use the term "copyright" but instead defines the specific right, the same approach of enumerating a power also eventually adopted for granting the power of "patent" rights for inventions, and exactly the approach of the Statute of Monopolies for permitting exceptions of allowed powers in a general policy of denial of power.[16]

That these various powers were considered and rejected suggests that the federal power to create, guarantee, promote, or limit economic rights was intentionally granted only in a very narrow form, that of the grant of limited-term exclusive rights to writers and inventors.

Our Modern "Government of Patentees"

How widespread is this dubious federal exercise of nonexistent general patent powers? Any federal agency that grants a license, certificate, franchise, or exclusive territory is potentially in violation. In any particular example it would require a careful review of the specific certificate or license power, and of any legislative restrictions

---

[14]The National Science Foundation Act and similar statutes therefore are also probably unconstitutional.

[15]See Cohen, *Science and the Founding Fathers,* as well as Madison's *Notes* for August 18, 1787.

[16]Other options considered and rejected were the power to establish "seminaries" (places of study), "public institutions, rewards, and immunities" for "promotion" of various purposes.

on its use, to determine if an unconstitutional patent was being granted. For example, federal licenses for operation of atomic power facilities could be acceptable under the present arguments, because they are nonexclusive in the relevant economic sense.[17] The license is to operate based on technical criteria, not to control a service territory.

It is very likely, however, that the grant of a local telephone (or other telecommunications service) monopoly territory by a state is unconstitutional—or at least voidable by the federal government under the commerce clause. Just as state laws could not be used to prevent steamboats from competing with sails (as discussed in an 1824 case, *Gibbons v. Ogden*, cited later in this chapter), state laws cannot prevent satellites and so forth, acting under federal authority, from competing with copper wires. Similar arguments also apply to cable services and telecommunications services generally.

In a 1991 decision known as "video-dial tone," the Federal Communications Commission applied similar logic in permitting telephone companies to provide video services over telephone lines. Since local cable television operators often have locally franchised monopolies for provision of cable television, this places the telephone company in direct competition with the cable company despite the local franchise. When challenged in court by cable operators seeking to protect markets and by local franchising authorities fearing loss of tax revenues, the appeals court upheld the federal action. Although the case has not (yet) challenged the constitutionality of such local franchises given the obvious nonexistence of a natural monopoly for television services, this court decision, so far as it goes, is the correct legal result.[18]

The Energy Policy Act of 1992 may be unconstitutional. The Act imposes federal planning requirements on states that presume the use of the state's general patent power.[19] The Supreme Court has dealt cleanly with this issue:

[17] 42 USC Section 2133(b).

[18] See "Phone Firms Exempted from Cable Franchise Fees," pp. F1 and F8 in *The Washington Post*, August 27, 1994.

[19] This statement is not contradicted by the findings in *Federal Energy Regulatory Commission v. Mississippi*, 456 U.S. 742 (1982). In that case, in the Public Utilities Regulatory Policies Act of 1978, Congress had required that states, including Mississippi, hold hearings to consider certain regulatory standards for electric utilities. When Mississippi challenged the requirement, the Supreme Court found that if no particular outcome were mandated to occur as a result of such hearing, the discretion

> Congress may not simply commandee[r] the legislative pro-
> cesses of the States by directly compelling them to enact and
> enforce a federal regulatory program.[20]

and:

> The allocation of power contained in the Commerce Clause,
> for example, authorizes Congress to regulate interstate com-
> merce directly; it does not authorize Congress to regulate
> state governments' regulation of interstate commerce.[21]

Mandatory "alliances" for medical care services as proposed in
the 1994 Clinton Health Care Initiatives would have been unconstitutional because they would have granted general patents that the
federal government has no power to grant. The proposal would
have, in effect, forced providers to sell services only to a federally
franchised monopsony buyer, while forcing individuals to buy services only from the same entities, as franchised monopoly sellers.[22]
There can be no more clear example of a prohibited exercise of the
nonexistent federal general patent power. Nor, because of the Tenth
Amendment, can the federal government require the states to form
such entities, because to do so requires exercise of a power, the
general patent power, that was clearly reserved to the states.

Federal regulatory authority deriving from the use of certificates
has been challenged frequently on other grounds, resulting in the
present framework of restrictions on procedure. For example, some
have challenged the application of regulatory authority as an unconstitutional "taking" prohibited by language in the Fifth Amendment
(and of the similar language of the Fourteenth as applied to the

---

of the state was not violated. In more recent legislation, such as the Energy Policy
Act, which requires certain forms of planning, and similarly the Clean Air Act Amendments, the transgression into actual mandates of particular action by states is more
clear, and therefore probably prohibited even under the *FERC v. Mississippi* doctrine.
Further, a recent finding in Wyoming Federal District Court, invalidating part of the
Brady Bill on gun control as violating the Tenth Amendment, implies that federal
mandates that require a state to use state resources on a matter otherwise reserved
to the states by the Constitution, may also be unconstitutional.

[20]See Hodel v. Virginia Surface Mining & Reclamation Assn., Inc., 452 U.S. 264,
288 (1981).

[21]See *New York v. United States*, 488 U.S. 1041, electronic version, 1992.

[22]For an economic analysis of the Clinton Health Care Plan, see a symposium in
the Summer 1994 issue of *The Journal of Economic Perspectives*.

states) to the Constitution: ". . . nor shall private property be taken for public use without just compensation." The courts denied these challenges, holding that price regulation of utilities' holding certificates was not a prohibited taking, so long as certain conditions on pricing were met. In one case the company was operating under a federal certificate granted under the Natural Gas Act and therefore, by the present theory, was regulated pursuant to an unconstitutional federal general patent. But general patent power questions were neither raised nor decided in the classic cases, as reviewed in this book.[23]

General Patent Power and Antitrust

Does the nonexistence of a federal general patent power also prohibit exercise of federal antitrust power? We shall not examine the question in depth here because this book is about energy policy, not federal regulatory policy in the more general sense. However, some comment is due and the considerations would be similar. For example, because the purpose of the federal commerce clause is, in large part, to prohibit states from interfering with commerce, antitrust power directed against state action would likely be constitutional.

The proper limit of federal antitrust power, therefore, lies somewhere between allowing federal antitrust regulation that limits the

---

[23]The Supreme Court decisions in the *Hope* case and the *Market Street Railway* case are discussed in more detail subsequently. But a comment is appropriate here. In *Hope*, and more specifically in the *Market Street Railway* case, it was also found that the obligation to assure "just compensation" under the takings clause did not extend to assuring revenues when market forces made the products obsolete or the prices too high. That is, neither the takings clause nor any exercise of regulatory authority over pricing was intended to substitute for competition or open markets. This is true for state or federal authority, since the *Hope* case involved authority of the Federal Power Commission while *Market Street* involved California state regulatory authority. These rulings should give pause to those who suggest that electric utilities should be compensated if the transition to a competitive market leads to an economic "stranding" of their assets. Besides being unwise (see Louis Kaplow, "An Economic Analysis of Legal Transitions," *Harvard Law Review*, vol. 99, 1986, pp. 511–617, and Richard Sansing and Peter M. VanDoren, "Escaping the Transitional Gains Trap," *Journal of Policy Analysis and Management* 13, Summer 1994, pp. 565–70), compensation may be unconstitutional as well, since it involves federal exercise of the general patent power. The *Market Street* case, in any event, defeats any argument that utilities have an expectation of recovery of costs of obsolete equipment.

ability of states to interfere with interstate commerce, and prohibiting it when the federal antitrust action amounts simply to allocating a market. For example, antitrust ought not be used merely because the federal authority thinks that it is "good policy" to break up a market participant that is "too big."

Recent federal antitrust action in the semiconductor industry illustrates the issues involved in determining the proper boundary. The grant of a federal copyright or patent—that is, of a limited-term exclusive right—to a software company or manufacturer might also result in the holder of that copyright or patent dominating markets for products similar to or dependent upon the licensed item. If the federal government then intervenes under antitrust laws to limit the market for the holder of the copyright or patent, it has overreached the specific patent power granted by the Constitution. Instead, it has used the general patent power allocating a market, a power the federal government lacks.

Such action also would contravene the purpose of the specific patent and copyright power that the federal government does have: to grant exclusive term-limited rights to authors and inventors to promote science and the useful arts. Taking constitutional language at face value, under this power the Congress may determine the term (time period) of exercise of exclusive use. Congress may change the length of the term in subsequent law for subsequent copyrights or patents. But federal agencies may not complain about private actions resulting from the owner of the right successfully exercising it for the very purpose for which constitutionally intended: allowing authors and inventors to profit from their creativity.

Sometimes, indeed fairly often, the result of competent invention is creation of useful commerce, which, in turn, makes large companies. If federal agencies act under antitrust to break up large companies merely because the government thinks the company or its market share is "too big," that is a threat not only to economic rights (such as proper exercise of copyright or patent) but also to political liberty.

Federal Patent Power: A Summary

The federal government may not grant exclusive general patents. No enumerated federal power exists to regulate business "affected with a public interest." No federal power exists to regulate merely because the absence of such a federal power would create a "vacuum" of regulation in interstate commerce. The federal government does not have powers just because the states lack them. Congress cannot create or exercise power not delegated to it by the federal Constitution merely by passing a law that purports to exercise a power that fills that vacuum.

The federal commerce clause restricts state exercise of their general patent power,[24] so long as no uncompensated takings occur, provided that the purpose of the regulation is to sustain open commerce among the several states, provided that the obstruction to commerce arose through state action, and provided that the federal action itself is not simply an allocation of markets. The federal government already does this, of course, through antitrust laws. Thus the federal antitrust laws are valid, despite the absence of a federal general patent power, to the extent that the purpose and application of the laws are to assure competitive interstate commerce when state action obstructs.

My analysis of the limits to federal regulatory power is thus also much more fundamental than the criticism of most critics of the New Deal Progressives. Critics of the Progressives debate the proper application of federal regulatory power, whereas I question the existence of much of that power.

States and the Patent Power: *Munn v. Illinois*

The historical facts related to the limits on the English Crown, which were the foundation for the American constitutional theory of limits on federal powers, have apparently been forgotten in the course of subsequent development of federal regulatory power. But what about the development of regulatory power by the states? The states regularly issued various general patents typically called licenses, certificates, or franchises.[25]

---

[24]This much at least was decided by the Supreme Court in one of the first antitrust law decisions, *United States v. E.C. Knight Company*, 156 U.S. 1 (1895). The literature on antitrust law generally is vast, as is also the literature on due process law following the Munn decision.

[25]The Supreme Court upheld this power, especially in the case known as *The Slaughterhouse Cases*, 83 U.S. 36 (1872). The relevant case for present analysis, however, is *Munn v. Illinois* of 1877, discussed later. The significant issue raised here but overlooked by other analysts is that both cases dealt with state, not federal, power and, therefore, the conclusion of either that the state could exercise certain general patent powers has no application to whether the federal government could exercise similar powers.

"Affected with a Public Interest"

By the late 19th century, a particularly vexing problem for state legislators was popular complaints about some rather peculiar new services and technologies. People would move west, carve farms from a forest or prairie, and start to raise grains. Then, others would follow and build places to store the grain (called elevators), and perhaps a railroad to carry the grain to market.

Not surprisingly, the elevators and railroads quickly became vitally important parts of the commercial landscape. Because consumers frequently encountered few choices about the railroads or elevators with which they could do business, both industries became the focus of citizen anger. For example, the business practices of the Munn & Scott elevator company of Chicago made no friends among a then-powerful political movement, the Grange, one of whose political objectives was price control for railroads and other businesses serving the prairie farmers.[26]

One public accusation made against Munn and Scott is that they sold receipts for grain not physically present in their elevators. Even though the practice probably appeared outrageous in the 1870s, today a similar practice, "selling short," is commonly used in both commodity and stock exchanges as a risk reduction technique. As long as the seller of such contracts makes good on the obligation by time of delivery (including the possibility of repurchase of the contract), the existence of the practice enhances efficiency. Such practices of Munn & Scott's elevator company would today not even fall under the jurisdiction of modern utility regulatory commissions.[27]

What could be done about citizens' complaints? Requests for federal intervention would do no good, since the federal government apparently lacked the direct power to affect such commerce, especially as viewed in 1870. But the state legislatures were not so confined, and indeed state legislatures directly regulated elevators, especially their prices.

[26]The historical context of the case is reviewed in an article titled "A Foot in the Door" by C. Peter Magrath, *American Heritage*, February 1964, vol. XV, no. 2, pp. 44–48, 88–92. A similar historical view of the political context of *Munn* is presented in Hofstadter and Hofstadter, *Great Issues in American History, Volume III, From Reconstruction to the Present Day, 1864–1981*, rev. ed. (New York: Random House, 1982). That text also reprints a portion of the *Munn* decision, and a portion of an influential dissent by U.S. Supreme Court Associate Justice Stephen J. Field.

[27]The worst accusations against Mr. Munn would probably amount today to fraud

Not surprisingly, the operators of such storage devices did not like regulation of their businesses, and they sued, or more accurately, appealed their conviction for failure to comply with the state regulation.[28] The complaint of elevator operator Ira Y. Munn against the state of Illinois reached the Supreme Court of the United States in 1877.[29] Munn lost. The Supreme Court found that the operation of a grain elevator was a business "affected with a public interest." Therefore it was within the power of the state of Illinois to regulate its right to do business and its prices.[30]

The Court implied that the notion was simply an extension of existing English common law under which the Parliament could regulate a business "affected with a public interest." The Court stated that if such operator were unhappy with regulation he "... may withdraw his grant by discontinuing his use; but so long as he maintains the use, he must submit to the control."[31] The "grant" in question here is the offer by Munn of the use of his private

in the sale of negotiable instruments, and if proven would have civil remedy based on both contract law and perhaps some tort claims. Other accusations, such as the charge that Munn and other major Chicago elevator operators conspired in their business practices, would today be seen as antitrust violations. Moreover, such practices would also be subject to various forms of remedy apart from those available from utility commissions. The fact that the allegations against Munn would today fall under other identifiable areas of law strengthens the argument that present antitrust law should preempt state power of utility price regulation coupled with the state grant of monopoly certificates.

[28]Many Supreme Court cases and a large body of analytical literature can be found on the powers of states. Only the classic cases that are used to support regulation of the energy industry are discussed.

[29]In the case *Munn V. Illinois*, 94 U.S. 113 (1877). Munn and his partner George L. Scott were appealing to the federal courts from a state conviction under regulations allowed by the constitution of the state of Illinois. Rather detailed discussion of the role of the *Munn* case in the federal constitutional law affecting the extent of state regulatory jurisdiction is found in Chapter VII of Charles Fairman, *Reconstruction and Reunion 1864–88*, Vol. VII, part 2 of the Oliver Wendell Holmes Devise, *History of the Supreme Court of the United States* (New York: Macmillan, 1987).

[30]*Munn* also initiated the two-part conceptual framework of regulation, the granting of a "certificate," and the regulation of price for services offered under that certificate. Since such "certificates" clearly are government grants of rights, and since they were and still are often granted as exclusive rights, this practice is clearly a derivative of the general patent power.

[31]95 U.S. at 126.

facilities to the public. The Court decided that this combination of licensing of entry and price control power was constitutional when exercised by the states.[32]

Once the state power to regulate businesses "affected with a public interest" was affirmed, it wasn't too difficult to find other businesses similarly affected. Constituent demands made it hard for legislatures not to regulate businesses "affected with a public interest."[33]

The Birth of Utility Commissions

Finding a just and reasonable utility price was a difficult job for a legislature. Given constituent pressures leading to such regulation, it seemed also an unavoidable one. But the job turned out to be not entirely unavoidable, since politicians are very good at one activity: passing the buck, and, when possible, passing the buck to a committee. And that, in brief, is the function of a state regulatory commission: a special committee created by the legislature to which it delegates difficult, if not impossible, economic tasks. This committee is also delegated certain legislative powers, in particular to find facts (hold hearings), to set prices, and often to carry out other state legislatively assumed duties such as the grant of exclusive territories—which are of course state general patents, also often called

---

[32]Reading the dissent of Justice Field in the *Munn* case in particular makes clear that issues covered in *Munn* were about state powers. The case dealt with federal powers only to the extent, first, of analyzing whether application of the federal Fifth Amendment "takings" clause to price and licensing regulation resulted in an improper taking of private property by the state of Illinois (Field thought it was); and second, whether the state made proper application of the Fourteenth Amendment right of due process in the course of such takings. The history of constitutional case law following *Munn*, and much else as applied especially to antitrust law and regulatory law generally, deals only with these two issues. For reviews of the literature that supports the view of the present text that the cases have discussed primarily process, see such texts as Edward Keynes, *Liberty, Property, and Privacy, Toward a Jurisprudence of Substantive Due Process* (University Park, Pa.: Pennsylvania State University Press, 1996); or John H. Rohr, *To Run a Constitution, The Legitimacy of the Administrative State* (Lawrence: The University Press of Kansas, 1986); or Hadley Arkes, *The Return of George Sutherland, Restoring a Jurisprudence of Natural Rights* (Princeton, N.J.: Princeton, University Press, 1994).

[33]Railroads, whose activities came under the control of specially designated state agencies called "Commissions," were the most frequent targets. The Texas Railroad Commission, which today regulates oil, gas, and utilities, derived its name and original powers in this manner.

franchises. Once created, however, the regulatory commissions found that their ability to legislate was confined.

As a result of due process and property-takings considerations resulting from the Fifth and Fourteenth Amendments, court enforcement of the limits on government required regulatory actions to be based on facts. Thus, the commissions were forced to rely more and more on evidence. And because they were regulating operating business, evidence was necessary on the engineering, accounting, and economic realities of the business regulated. By this very circuitous route, the regulatory commissions, whose original purpose was to solve a political problem for state legislatures, now are seen by the public as serving primarily engineering, economic, and accounting purposes.

The interaction of political purpose with economic means pervades the history of energy regulation. This occurs for several related reasons. First, the American constitutional and legal system makes it difficult for government to act outside certain prescribed areas. Because politicians have not seen fit to limit their desires to act within those boundaries, they have had to find creative ways to meet their purposes while seeming to remain within legal limits.

Second, the administrators who carry out the laws do not themselves always remain within their bounds. For example, the following views are attributed to James Landis,[34] a chairman of the federal Securities and Exchange Commission in the New Deal era:

> One of the ablest administrators . . . never read, at least more than casually, the statutes he translated into reality. He assumed that they gave him power to deal with the broad problems of an industry, and upon that understanding he sought his own solutions.

Indeed, there is ample evidence that regulators historically have acted (and been implicitly allowed to act) far outside the realm of their delegated powers. After all, without the checks on lawmaking

---

[34]Stated as a quotation but without citation as to source in a book review by Administrative Law Judge Donald W. Frenzen, in his book review at pp. 74–76 in the *Federal Bar News and Journal*, January 1994, vol. 41, no. 1, of the book Morton J. Horwitz, *The Transformation of American Law, 1870–1960: The Crisis of Legal Orthodoxy* (New York: Oxford University Press, 1992).

power that exist in legislative bodies (such as executive veto authority, filibuster options, committee jurisdiction, etc.) regulatory findings are not as easily blocked by opponents even if they are in the minority (which can often exercise significant legislative power).

All the rules of order and procedures that the founders established for legislative activities were designed to ensure a large degree of political consensus before rules can be adopted. That is not to say that the regulatory consensus would not carry the day on the legislative floor, but once that power has been delegated to commissions the only "consensus" that must be reached is that of the members of the commission itself. The members often are politically appointed individuals who do not represent the public as a whole. Moreover, the decisions are made outside the hot spotlight that usually is centered on the legislative floor or governor's mansion. Thus, the typical utility rate structure might contain subsidies of suburban and rural electricity customers at the expense of urban customers, higher rates to industrial customers than are justified by cost, and so forth. Those two realities among others set up a situation in which those with political agendas on the commissions are given a freer than realized hand to pursue those agendas.[35]

Despite the enormous implications of legislatures delegating important powers to regulatory commissions,[36] the legal issues brought to the courts after *Munn* were about more mundane matters: procedure (such as the conditions under which particular regulatory rate decisions reflected due process), or "takings" (whether particular regulations took private property for public use without just

---

[35]This issue was called to the author's attention by Jerry Taylor of the Cato Institute. It is not a standard criticism made against regulation by delegation, perhaps because we have become so accustomed to the practice. But it is a nonetheless valid one. To understand the reality of the risks of excess posed by such delegation, review for example the many discussions of the "proper allocations" of state and federal power, given by the Federal Energy Regulatory Commission in its recent Order 888, in April 1996. Being given a free hand in legislation by delegation is no longer even enough; now these bodies write their own theory of federalism and act on their own view of how to rewrite the Constitution. See David Schoenbrod, *Power Without Responsibility: How Congress Abuses the People Through Delegation* (New Haven: Yale University Press, 1993).

[36]For classic statements of similar concerns of political scientists see Theodore Lowi, *The End of Liberalism* (New York: Norton, 1969) and Grant McConnell, *Private Power and American Democracy* (New York: Knopf, 1966).

compensation).[37] The question of whether the federal government had the power to regulate market entry, allocate market shares, and set rates at all—and especially whether it had power to grant exclusive territories in the same manner as states—was conspicuous for its absence.[38]

The Commerce Clause: A Restraint on State Patent Power

Does Illinois possess the (general patent) power to regulate grain elevators? The Tenth Amendment of the Constitution says that "The powers not delegated to the United States by the Constitution, nor prohibited by it to the States, are reserved to the States respectively, or to the people." The general patent power was not one of the powers either prohibited to the states or reserved (except in a specific form) to the United States. Therefore, the people of the state of Illinois can empower the state of Illinois to exercise such general patent power if they wish (and so long as that power is not otherwise preempted by some other federal action, which, in 1877 in the *Munn* case, it was not).

What does proper federal exercise of the commerce clause say about exercise of the state general patent power? States may both

[37]Two famous sets of cases in the Supreme Court's October term in 1912 show the detail of consideration needed to set rates that also took account of business realities, not merely political ones. The 1912 decision in *The Minnesota Rate Cases* (230 U.S. 352) concerned complaints about how the state of Minnesota set railroad rates. This decision not only laid out the requirement that rate-making decisions be factually based—the criteria for this are explicitly stated on pp. 434–35—but the tedious detail of the decision even includes analysis of operating expenses of several railroads. On the same day, both the state of Missouri and numerous railroads complaining about its regulations learned the same detailed lessons in a group of decisions known as *The Missouri Rate Cases* (230 U.S. 474).

[38]A thorough review of the legal development of both state and federal regulatory powers following *Munn* is given in Owen M. Fiss's *Troubled Beginnings of the Modern State, 1888–1910*, part of a series of studies that reviews all Supreme Court cases (Volume VIII of the Oliver Wendell Holmes Devise publication *History of the United States Supreme Court* (New York: Macmillan, 1993). The work devotes an entire chapter to the topic "Rate Regulation: The Assault on *Munn v. Illinois*," but also considers related subjects such as the rise of antitrust regulation and the 1906 *Hepburn Act*. The book pays great attention to details about the proper limits of regulatory agencies and their relationships to property rights. Once the 1877 *Munn* decision found the existence of a plenary state power to regulate common carriers "affected with a public interest," no one seems to have further asked whether there existed a corresponding federal, enumerated, power.

96

grant certificates (as exclusive patents or "franchises" in the modern terminology) and regulate prices so long as they do not commit uncompensated takings, but states also are not required by the takings clause to act contrary to ordinary behavior of markets. At the very least, the case law about the commerce clause to date indicates that a company does not obtain antitrust immunity merely because its prices and services are subject to state regulation.

Beyond this, the question of how the existence of exclusive state certificates affects interstate markets is an empirical one: do they restrict entry? Under the classical view—that utilities are natural monopolies—the answer might be "no," with the result that the states can continue to exercise their patent powers by granting exclusive territories. One might argue the exclusive grant does nothing to restrict a market that would not have occurred in any event.

But if utilities are *not* natural monopolies (and the evidence presented in Chapter 2 suggests that they are not), then exercise of the state patent power surely *does restrict* entry to a market that otherwise would have remained open. That power must, therefore, be preempted as an unacceptable restriction on interstate commerce. In fact, such state-granted monopolies appear to violate antitrust laws.

This interpretation of state patent power is relevant to a growing debate in law journals over the present meaning of *Munn*. Some authors seem to believe that regulation in the traditional manner is justified by *Munn* because utilities are purportedly natural monopolies.[39] Others argue that *Munn* justifies regulation despite the presence of natural monopoly and, indeed, even argue that *Munn* creates a "public interest exception" that permits or even requires regulation by either federal or state government despite the presence of "unfettered competition."[40]

In fact, neither side of the current academic "debate" is correct. Neither energy production, distribution, nor transmission is a natural monopoly; and in any event the Court found in *Munn* that states had plenary power to regulate common carriers based on the Court's

---

[39]See Douglas Gegax and Kenneth Nowotny, "Competition and the Electric Utility Industry: An Evaluation," in *Yale Journal on Regulation* 10, 1993. See also footnotes to traditional economics texts such as by Samuelson, cited in note 43 of Richard D. Cudahy, "Retail Wheeling: Is This Revolution Necessary?" *Energy Law Journal*, vol. 15, no. 2, 1994, pp. 351–63.

[40]Cudahy, pp. 360–61.

understanding of English common law, not on some appreciation of a "special exception" that permits states to regulate despite the presence of competition.

Notwithstanding *Munn*, the states are specifically limited by the federal Constitution from interference with interstate commerce, for which purpose the commerce clause grants legislative powers to Congress. At the time of *Munn*, antitrust laws did not exist, but today they do. Therefore, despite *Munn*, states cannot close markets that federal powers have opened. With or without natural monopoly, the effect of traditional state regulation is to close utility markets, but without natural monopoly, the only thing that keeps them closed is state regulation. Therefore, despite *Munn*, traditional state utility regulation is today unconstitutional as a preempted restraint on competition in interstate commerce. Further, since *Munn* dealt only with state power, that case cannot be a justification for federal exercise of restrictive powers of entry. The federal government has no general patent power, and *Munn* did not create it.

Gibbons v. Ogden: The Limits of State Regulatory Power

Is there a federal power that preempts the state use of exclusive grants?[41] Do federal antitrust laws, if constitutional, preempt state regulation that otherwise tends to close markets by the state grant of certificates (patents) under the powers reserved to the states?

Both questions are answered, in large part, by a Supreme Court decision, written by John Marshall in 1824, known as *Gibbons v. Ogden*.[42] The case would have been of interest to historians if for no other reason than that it involved state licenses for the use of steamboats operating between New York and New Jersey originally granted in part to Robert Fulton, the American inventor of steam-powered ships. By the time the dispute reached the United States Supreme Court, federal coastal licenses had also been issued for steamboats for certain purposes.

The issue was whether a federal license permitted the holder to operate in New York, without a New York license, and within the exclusive territory of the grant given by New York. In arguing for

---

[41]George Dangerfield, "The Steamboat's Charter of Freedom," *American Heritage*, October 1963, vol. XIV, no. 6, pp. 38–43 and 78–80.

[42]22 U.S. (9 Wheat.) 1, 6. L. Ed. 23 (1824).

Gibbons, attorney William Wirt posed questions related to the existence of a state patent power, similar to that defined for the federal Congress in Article I, section 8, paragraph 8.[43] The Court decided the case as follows:[44]

> This [federal] act demonstrates the opinion of Congress, that steamboats may be enrolled and licensed, in common with vessels using sails. They are, of course, entitled to the same privileges, and can be no more restrained from navigating waters and entering ports which are free to such vessels, than if they were wafted on their voyage by winds, instead of being propelled by agency of fire. The one element may be as legitimately used as the other, for every commercial purpose authorized by the laws of the Union; and the act of a state prohibiting the use of either to any vessel having a license under the act of Congress, comes, we think, in direct collision with that act.
>
> As this decides the cause, it is unnecessary to enter in an examination of that part of the constitution which empowers Congress to promote the progress of science and the useful arts.

Thus, the action of Congress preempted state licenses or patents because Congress issued licenses in a manner that opened markets among the several states. It was not even necessary for the Court to enquire further whether there was a limited state patent power similar to the limited federal patent power; the fact that Congress had exercised commerce power to open markets was sufficient. In 1877, more than 50 years later, the Court in *Munn v. Illinois* found for the existence of a state licensing authority, which implied a more general state patent power. At that time there was no conflicting federal regulation of the same subject matter and therefore no preemption.

The 1824 *Gibbons* decision is also a model for how to decide whether federal antitrust laws preempt state grants of utility certificates in the absence of monopoly. Modern analysts cite the case's

---

[43]For an interesting discussion of historical and other issues involving this case, the parties, and the attorneys, see discussion in Chapter VIII of G. Edward White, *The Marshall Court and Cultural Change 1815–1835*, abridged edition (New York: Oxford University Press, 1991).

[44]9 Wheat. at 221.

finding that the congressional power to regulate commerce "is the power to regulate; that is, to prescribe the rule by which commerce is to be governed."[45] Yet they often do not apply the analysis to the fact that the case related to voiding the effect of a grant of an exclusive territory by a state when no natural monopoly exists.

An exception to the present wide acceptance of use of *Gibbons v. Ogden* as the basis for modern expansive federal commerce powers is the analysis offered by Roger Pilon, director of the Cato Institute's Center for Constitutional Studies:

> Chief among those chinks [in the Constitution found by New Deal lawyers] was the Commerce Clause, which had been written, ironically, not to facilitate regulation but to enable Congress to override protectionist regulations that states had passed under the Articles of Confederation. Written thus to "make regular" commerce among the states—much as the court used it in the first great Commerce Clause case, *Gibbons v. Ogden*—the Clause was seen by Progressives as affording Congress the power to affirmatively regulate commerce for all manner of social ends.[46]

Both Pilon and I are suggesting that *Gibbons v. Ogden* be applied as originally written. The overextension of federal regulatory power predates the Democratic programs of the New Deal by a half century. The New Deal resulted in the expanded application of federal and state administrative powers, but did not create the basis for that application. Neither did *Gibbons v. Ogden* as originally written. Those powers were given birth by *Munn v. Illinois* and subsequent decisions built upon that decision, even though the *Munn* decision was about the power of states. Federal general patent power regulation occurred apparently through the introduction of regulation to federal territory.

---

[45]This specific text is contained within an extended selective citation from *Gibbons v. Ogden*, cited in the chapter on "Constitutionality of the Sherman Act," in one of the fundamental sources of the law of antitrust, Earl W. Kintner, *Federal Antitrust Law, A Treatise on the Antitrust Laws of the United States, Volume I: Economic Theory, Common Law, and an Introduction to the Sherman Act* (Cincinnati: Anderson Publishing, 1980). Citations from this 1824 case are the single most extensive discussion on the constitutionality of antitrust laws given in that text.

[46]"Does The Constitution Restrain Congress?" *The Washington Times*, November 8, 1994, p. A15.

Because territories are not states, a common law argument, such as in *Munn v. Illinois*, might permit the federal government to regulate business "affected with a public interest" by direct and exclusive grants within those territories. This seems to be the basis by which the Congress directly regulated railroads in federal territories in the 19th century and subsequently devolved that power to the states. Such federal power expires when the territory becomes a state. However, there seems to have been no case that considers the general patent power applications of federal regulation of market entry and price by commissions.[47]

Conclusion

Ira Munn is an unlikely but proper hero. He worried about the extension of the power of government beyond the boundaries permitted by the Constitution. Nobel Prize winner F. A. Hayek also worried about the extension of the power of the English Crown beyond that permitted by the Statute of Monopolies. Hayek cited from the words of Sir Edward Coke specifically on the relationship of an English common law decision related to that Statute, to the Magna Carta, the foundation document of English liberty,

> if a grant be made to any man, to have the sole making of cards, or the sole dealing with any other trade, that grant is against the liberty or the freedom of the subject, that before

---

[47]In 1906 the Congress passed *the Hepburn Act*, which permitted the Interstate Commerce Commission to regulate interstate oil pipelines in the same fashion as Illinois regulated grain elevators. In 1913 in a decision known as *The Pipe Line Cases*, 234 U.S. 548, the Supreme Court found that the powers of Congress to regulate interstate commerce extended to the power to regulate the manner in which a private business in interstate commerce could conduct that business.

More precisely, the Court found that *The Hepburn Act* was valid because the effect of its application was to prevent the pipeline in question from requiring sale of oil to the pipeline before the line would carry that oil. In essence, this says that because the Commission used an antitrust-type analysis related to market structure, and in the end opened a market, the result of that analysis was also valid within the federal commerce power. The Act, however, did not address the general patent power issues discussed here.

A much more complete history of the evolution of utility regulation, though with somewhat different emphasis, is found in Robert Bradley Jr., *Oil, Gas & Government: The U.S. Experience* (Lanham, Md.: Rowman & Littlefield, 1996).

did, or lawfully might have used that trade, and therefore is against this great charter.[48]

Hayek makes clear that the Statute of Monopolies was a critical part of the struggle to limit the power of the government in England. The Founding Fathers of the United States designed the Constitution with similar limits in mind. The bill of complaint against Ira Munn was that he refused to permit the government to limit his right to do business and to set his price. It is no crime to fail to do that which the government has no power to compel. If Hayek is right, and if Ira Munn operated a public utility in the United States today, he might win in the case of *Munn v. Illinois, II.*

---

[48]F. A. Hayek, *The Constitution of Liberty* (Chicago: University of Chicago Press, 1960), p. 168.

6. What Should Be Done?

We need to fix energy policy. And what we need to do is simple in concept but politically dramatic in practice: repeal all the state and federal laws that regulate energy, including those authorizing both state and federal utility regulatory commissions. The economic rationales used to justify such regulation are no longer supported by facts and never may have been supported by the facts. Given that political bodies are considering the deregulation of the electricity industry,[1] perhaps the time is ripe for a truly revolutionary agenda.[2] Why not remove such regulation entirely?

The Collapsing Case for Regulation

Why should we consider changing the regulatory status quo? First, the regulated monopolist provides energy at higher rates than the unregulated monopolist. An unregulated monopolist must face the threat of competitive entry, including the invention of new and better technologies than his own. The unregulated monopolist must face the reality that people have other uses for their own money. But regulated franchised monopoly is protected even from the threat of competition, must add the direct cost of regulation to its own unrestrained costs, and is subject to perverse regulatory schemes that merely add even more to the cost of the utility and the regulator.

[1]As of this writing, the Federal Energy Regulatory Commission has issued its Order 888, restructuring the service offerings of essentially all electric utilities. Most states are enforcing similar changes locally—the California Commission even maintains a Web site of cross references to other states and related actions on restructuring nationally. Congress has received bills from Representatives Ed Markey and Dan Schaeffer that would force states to certify whether local electricity markets are competitive and to remove traditional regulations. And in allied fields such as telecommunications, even local areas finally may have been opened to competition as a result of Congressional action.

[2]For a discussion of why conventional utility regulation is beginning to collapse, see Benjamin Zycher, "Market Deregulation of the Electric Utility Sector," *Regulation, the Cato Review of Business and Government*, Winter 1992, pp. 13–17.

Second, whenever competition is allowed in markets, prices have invariably fallen. Economist Paul MacAvoy, who began federal deregulatory efforts under the Ford administration, recently told the Federal Energy Bar Association that deregulation in the American transportation and natural gas industries has been followed by price reductions of one-fourth to one-third in the first four years. Recent studies by scholars from the National Regulatory Research Institute, the Brookings Institution, and the Center for Market Processes all confirm that deregulation has invariably increased market efficiency and reduced the price of goods and services.[3] In Argentina, deregulation of electricity has resulted in price reductions of as much as half.[4] In Lubbock, Texas, one of two competing electric power utilities was able to acquire new coal-fired generation capacity for under $300 per kilowatt, while the cost for similar capacity to traditional monopoly utilities nationally was about $750 per kilowatt.[5] The price of electricity in U.S. cities that have competitive supply is from 16 to 19 percent lower than in cities with monopoly supply.[6]

Stigler and Friedland studied rates of utilities in states before and after regulation and found that regulation did not lower rates, the purported public interest purpose used to justify regulation to the public.[7] A more recent study replicated the Stigler and Friedland study in the five states that added regulation in the 1960s and 1970s. The result:

> The most interesting conclusion is that state regulation has no significant effect on electricity rates. Also of interest is the finding that investor-owned utilities do not have significantly higher rates than publicly owned systems, in spite of the . . . important subsidies received by the latter. Apparently these

---

[3]See Robert Crandall and Jerry Ellig, "Economic Deregulation and Customer Choice: Lessons for the Electric Industry," Center for Market Processes, 1997, and Kenneth Costello and Robert Graniere, "The Deregulation Experience: Lessons for the Electric Power Industry," National Regulatory Research Institute, August 1996.

[4]See J. Friedland and B. A. Holden, "Power Structure, Utility Deregulation in Argentina Presages Possible U.S. Upheaval," *Wall Street Journal*, June 19, 1996, p. 1.

[5]Jan Bellamy, "2 Utilities Are Better Than One," *Reason*, October 1981, p. 26.

[6]Ibid.

[7]George Stigler and Claire Friedland, "What Can Regulators Regulate?" *Journal of Law and Economics* 5, October 1996, pp. 1–16.

subsidies were offset through inefficiencies associated with government ownership and management.[8]

The study also found that the cost of regulation was significant and measurable, and concluded:

> The results indicate that public utility rate regulation offers no net benefits to consumers and results in efficiency losses. given the fact that there are measurable costs of regulation . . . it follows that deregulation of this industry would produce net benefits for society.[9]

The authors of that study compared the predictions of the classical textbook model of the behavior of a monopolist to the prices created by regulation. They examined the data in four separate ways. Two of the methods showed no statistical difference at all between the prices of an unregulated monopolist (one actually showed a lower price from unregulated monopoly) and prices under regulation; while the other two showed no more than a 5 percent lower price from regulation. The authors, however, believe that their analysis actually overestimated costs and "Hence the monopoly price is probably too high" in their study; they therefore conclude, "regulation is having little effect on price."[10]

---

[8]Mike A. Denning and Walter J. Mead, "New Evidence on Benefits and Costs of Public Utility Rate Regulation," Chapter 2 in *Competition in Electricity: New Markets and New Structures*, J. Plummer and S. Troppmann, eds. (Arlington, Va.: Public Utilities Reports, n.d.), p. 28. The study also tested whether the A-J effect, discussed earlier, had a measured effect on capital costs, and found a statistically significant effect that regulated firms had a capital stock about 6.2 percent higher than nonregulated ones (p. 30).

[9]Ibid., p. 35.

[10]See pp. 372 and 374 from Thomas Gale Moore's "The Effectiveness of Regulation of Utility Prices," pp. 365–75 in *Southern Economic Journal*, vol. 36, no. 4, April 1970. Interestingly, the study also reviewed the "A-J-W" effect, which they refer to as the excess capacity problem, and found no evidence to support the view that regulation creates excess capacity. If regulation has no effect on utility investment (good or bad), and cannot cause prices lower than unregulated monopoly, what is the purpose of continuing regulation?

I am skeptical of the method used by Moore to estimate this effect. He compares generation plant investment for certain private utilities to generation investment costs of municipal ones. But municipal utilities faced peculiar incentives for generation investments when the data for the study were collected. Thus, municipals may have overinvested in plant, compared with competitive market utilities, for different reasons than private regulated utilities. If both municipal and private utilities overinvested, the comparison of municipals with regulated private utilities provides no information.

Going Cold Turkey

The Congress should repeal the charter of the Federal Energy Regulatory Commission, including the Federal Power Act, the Natural Gas Act and all subsequent amendments and subsidiary or related legislation including the Energy Policy Act, the Public Utility Regulatory Policies Act, and the Natural Gas Policy Act. The states should do the same.[11]

This book argues that the only role of the federal government in interstate commerce is to open markets closed by state action. Congress could do this for utilities by enacting legislation that would say, "Whereas any state action that has the effect of closing a market to interstate commerce may be removed by Congress, any state or territorial authority or affiliated commonwealth authority grant heretofore or hereafter made of any exclusive service territory, whether by charter, franchise, license, certificate, or other device is declared null and void."

Many will find the simplicity of this prescription disturbing. They might prefer that Congress impose some more elaborate comprehensive solution, because, many will say, the effects of deregulation will be felt among many states. Thus, no single state has the authority to take comprehensive action. The argument is true and irrelevant. Commerce across a state line always has affected more than one state, and any one state cannot unilaterally control such actions.

The Constitutional Convention considered and rejected language that would have given Congress power to act on matters in which a single state is incompetent to act. This is not because government should not act for the public good; it is because, if a government can act merely based on a claim of the public good, then a government can do anything without limit, because it is always possible to create a claim of public good to justify any action. The colonists knew that from their own, and England's, bitter history with the powers of an absolute king. Thus, the federal Constitutional Convention chose to give the national government only specific, enumerated powers, so that it could do only limited general good. Among those powers was the power to stop states from creating general harm

[11]And if this does not occur, some enterprising market-oriented utility company could achieve the same result in the courts by carrying out the agenda described in this chapter.

when they chose to act based on pretenses of the "general good." In any event, because the Congress has only the power to open markets closed by state action, it should do no more, despite claims about the general good asserted by those seeking to expand federal government control.

Advocates of uniform federal laws for the "public good" also believe that states are not sufficiently competent to draft their own laws for their own purposes. The evidence does not support this view. For example, we know that when uniformity among state laws has commercial benefit, states are exceedingly capable of drafting, on their own initiative, suitable laws. For example, the Uniform Commercial Code, regulating the important commercial law of sales of goods and transactions in negotiable instruments, was drafted entirely by interests other than the federal government and then enacted by most states in substantially similar form without federal mandates.

The Case against Regulatory "Reform"

Why remove traditional regulatory authority entirely? Why not merely add the authority to hear antitrust questions to the repertoire of state or federal utility commissions? First, existing price and certificate regulation may well be unconstitutional. Given the inclinations of certain members of the U.S. Supreme Court, traditional regulatory mechanisms may soon be off the table, so to speak, for policy purposes.

Second, absent removal of the power to regulate, it is unlikely that commissions will voluntarily surrender that authority. Indeed, the most basic instinct of regulators is to regulate more, not less, and the outbreak of competition in gas and electric markets may only appear as a complicating nuisance to be overcome by still more comprehensive regulation. Certainly, one explanation of the rise of state and federal requirements that utilities perform "integrated least-cost planning" at the same time that markets became more competitive is that regulators saw the ability to control markets slipping away because of increasing competition in the industry. They reacted by expanding their reach, rather than by limiting involvement.

The Perils of Political *"Restructuring"*

An example of government's tendency to use failure of past regulation to create new regulation is a bill recently introduced in the

Congress, known as the "Electric Consumers, Power to Choose Act of 1996."[12] The bill follows some of the prescriptions offered in this book, such as the repeal of the Holding Company Act. But in every instance, the removal of one form of regulation is predicated upon the creation of some new form, and sometimes of the perpetuation of the old form. For example, Section 104 of the bill deals with purportedly nonregulated utilities. It requires them to offer services that are "just, reasonable, and not unduly discriminatory and that permit the recovery by the utility of all costs incurred in connection with local distribution services and necessary associated services."

This seemingly innocuous language (innocuous to one not familiar with utility statutes) is exactly the language that presently governs state utility regulation. Enforcement of such language in the absence of a regulatory commission, that is, through a court, might be even more obstreperous than via a commission, as at present. As a practical matter, the presence of such language requires application of existing regulation to nonregulated utilities. So, nothing "deregulatory" at all would occur as a result of the bill.[13]

Worse, the proposed bill also simply appropriates for the federal government the state general patent power. Thus, if a state fails to make certain elections acceding to Congress's presumptions of how markets would be structured, then under Section 106 of the bill the Federal Energy Regulatory Commission (FERC) "shall exercise the authorities that would otherwise be exercised by state regulatory authority or nonregulated electric utility. . . ."

The federal government would simply assume the states' regulatory authority, an action probably not undertaken since the Reconstruction era following the Civil War, and then exercised only until the defeated states were reorganized. That this appropriation of power also would be an exercise of nonexistent federal general patent powers hardly needs further repetition in this book. It is also an exercise of the power specifically *not* granted the Congress by

---

[12]H.R. 3790, submitted by Rep. Daniel Schaefer, July 11, 1996.

[13]Like previous federal energy regulation such as the Holding Company Act, the bill also presumes to know the best form of the resulting market. It prohibits "nonregulated" utilities from "using any revenue from local distribution facilities to subsidize retail electric services provided by such utility. . . ." One can imagine the court and regulatory cases attempting to define "subsidize" for the purpose of the "deregulation"!

the Constitutional Convention, to act on any belief of the "public good" when a single state may be incompetent to act. Indeed, it is worse: it forces the state to act on the Congress's perception of the public good in an area of state competence, and then removes that authority if the state fails to act or disagrees. The ease with which the bill asserts federal powers that have no constitutional foundation is, or ought to be, frightening.[14]

The Perils of Regulatory "Restructuring"

A third reason exists why it is unwise to simply add pro-market language to existing statutes while giving the regulator the option to regulate under traditional or market-oriented powers: it has been tried and it does not work. It is true that some regulators, such as FERC, have been rather aggressive in permitting or encouraging competition within certain bounds. But it is by no means clear that that direction was freely chosen by Commission regulators. The Commission's 1985 "deregulatory" action, Order 436, actually increased the amount of regulation in the industry. The 1992 action advertised as "deregulation," Order 636, is more market-oriented, but leaves us with nothing like an unregulated market.

All manner of price details, and especially the right of market entry, are still tightly controlled by the regulator. Indeed, the Commission is now debating whether to control prices for gathering natural gas in producing areas, a subject traditionally outside its jurisdiction.[15] It is arguable that the primary reason that the FERC opened entry to the degree that it did was because Congress directly or indirectly required FERC to do so under the Natural Gas Policy Act of 1978, in the Natural Gas Wellhead Decontrol Act of 1989, and in the Energy Policy Act of 1992.

Even when regulators are faced with the presence of competition, they generally prefer to cling to regulation. A recent example can be found in the behavior of the Federal Communications Commission (FCC). In the Cable Act of 1992, Congress granted the FCC the power to regulate rates of local cable television operators,

[14]See also the analysis of the bill in Jerry Taylor, "Electric Utility Reform: Shock Therapy or Managed Competition?" in *Regulation*, September 1996, pp. 63–76.

[15]For a review of the problem of expansion of federal power into regulation of natural gas gathering, see Cody L. Graves and Maria Mercedes Seidler, "The Regulation of Gathering in a Federal System," *Energy Law Journal*, vol. 15, no. 2, 1994, pp. 405–25.

but only in markets where there was no competition. The FCC study undertook to discover whether competition exists anywhere in the United States between cable operators and other sources of television. Notwithstanding the existence of almost universal access to free over-the-air television, in 1994 the FCC found instead that virtually the entire United States was within the grip of cable monopolies. Faced with a choice between traditional regulation and open markets, the FCC chose tradition.

But technology does not respect the decisions of regulators. Since that decision in early 1994, both RCA and Radio Shack have begun marketing home TV satellite receivers for under $400 (down from $900 at the time of first draft of this text) retail, plus a monthly access fee. Given "basic" cable television rates of about $25 to $50 monthly plus additional access fees for added channels, anyone can now replace their cable hookup with a capital cost "payback" period of two years or less. Even "monopoly cable services on natural monopoly right-of-ways" are subject to competition from satellites, as well as from free over-the-air television. Given the history of rapid declines in prices for consumer electronics, it is likely this cost will rapidly drop even further.

A strange monopoly it is that can be challenged by something found at falling prices at every retail shopping center in the country! As if to emphasize the point of this book, in the fall of 1994 the FCC recognized the presence of the alternative technology in a press release discussing its decision to continue regulating. Said the Commission, "DBS (direct broadcast satellite), merely a vision in 1990, is now available and is being rolled out around the country. Press reports suggest that initial demand for equipment necessary to receive the service has exceeded supply."[16]

While federal regulation of the cable industry is too new to blame for a shortage, local regulation—including restrictions on market entry—has been pervasive. Further, the price method applied by the FCC to national cable regulation has been the same as that applied by the Federal Power Commission to natural gas from the 1950s to the late 1970s: uniform prices or rates of return based on presumed industry standards. Thus, the condition of the American

---

[16]"First Annual Report to Congress on Cable Competition Adopted," Federal Communications Commission, press release No. DC-2653, September 19, 1994, p. 2.

cable television industry is very much like that of the energy industry: price and entry regulation in the presence of competition cause shortages of supply and delay the entry of more efficient technology.

The clear lesson from these examples is that less regulation is better. Restricting the ability of regulators to do no more than assure open markets, especially when states for their own local political reasons prefer to close them, is a far better policy. In any event, the proper constitutional role of the federal government under the commerce clause is simply to open markets closed by state regulation.

The Antitrust Option

What happens if politicians do not repeal all specialized energy regulation? What is the next best agenda? The answer is relatively simple: remove all the laws and regulations from the books. To the extent necessary politically, preserve the various state and federal energy commissions, but restrict their authority to little more than specialized antitrust courts for energy matters.

Existing antitrust laws prohibit practices that interfere with competitive markets. Antitrust is enforced by government action—the Federal Trade Commission, the Justice Department, and their state equivalents when acting on state antitrust laws—and also by private parties, who are permitted to bring a civil action for harm believed to result from certain anti-competitive practices. Courts in the United States regularly hear complaints about competitive practices. Thus, my "backup" proposal would do no more than place energy under essentially the same laws as the rest of the economy.[17]

---

[17]My antitrust position, however, does not imply that I endorse vigorous antitrust enforcement against energy markets. The enactment of antitrust laws at the turn of the 20th century was supported by the same political forces that advocated economic regulation in general. Just as regulation was often introduced first in markets in which it was not necessary (like competitive electricity markets), antitrust laws were often used against industries that were quite competitive. A recent study of the industries accused of monopolistic behavior prior to the passage of the first antitrust law in 1890 argues that ". . . industries that were accused of being monopolized in the late 1880's were in fact cutting prices and expanding output faster than the rest of the economy." See Thomas J. DiLorenzo, "The Origins of Antitrust: An Interest Group Perspective," *International Journal of Law and Economics*, vol. 5, pp. 73–90, 1985; and Thomas J. DiLorenzo, Donald J. Boudreaux, and Steven Parker, "Antitrust before the Sherman Act," in Fred S. McChesney and William F. Shughart II, *The Causes and Consequences of Antitrust, The Public Choice Perspective* (Chicago: University of Chicago Press, 1995), pp. 255–70.

The Architecture of Modern Antitrust

Section 1 of the Sherman Act made illegal "every contract, combination in the form of trust or otherwise, or conspiracy, in restraint of trade or commerce among the several States, or with foreign nations. . . ."[18] Section 2 of the Sherman Act further made illegal the behavior of seeking to monopolize trade. It did so in very general language: "Every person who shall monopolize or attempt to monopolize . . . shall be guilty of a felony." But what is an "attempt to monopolize"?

By 1914, Congress found it necessary to be more specific about exactly what kinds of behavior were prohibited by the antitrust laws. Sections 2 and 3 of the Clayton Act, therefore, defined prohibited actions in more detail. One could not "discriminate in price between different purchases of commodities of like grade and quality," except that "nothing herein contained shall prevent differentials which make only for due allowance for differences in cost of manufacture, sale, or delivery resulting from the different methods or quantities in which such commodities are to such purchasers sold or delivered." Normal cost-based differences among products and services cannot be taken as evidence of illegal price discrimination.

The need to distinguish between cost-based and discriminatory price differences is also a fundamental issue in regulatory proceedings. Public utility commissions must decide the price for various classes of service and types of customer, such as residential, commercial, perhaps large and small industrial; whether the customer takes a full package of delivered energy or only transportation; whether all energy storage and transmission services are to be separated and priced; and so forth. Regulators decide those issues based largely on evidence related to the cost required to provide for each type of service or each class of customer.

Thus, on this key activity of utility regulation—setting price for services—the essential difference between the existing regulatory regime and the antitrust orientation that I propose is a redefinition of the authority of regulators. Commissions would no longer set prices. Instead, they would assure the absence of illegal price discrimination as measured by the differences of prices among various

---

[18]Note that Section 1 prohibits the organizational tools used to create undesired monopolies: the contracts or trusts that codify economic relationships.

classes and types of service. Commissions might see very much the same kinds of evidence on costs and business practices as they do now. But rather than use that information to set prices, they would use it only to identify the existence of undue price discrimination.

For example, confronted with evidence that a local monopoly exists in the delivery of energy, the antitrust remedy might be to make the monopolist a common carrier, required to carry for all what it delivers on its own account.[19] The Clayton Act and traditional utility regulation have many parallels, but under traditional regulation the service territory of a utility company is protected, by the authority of a utility regulator, from competitive entry. Under an antitrust regime, however, the goal is to open the market, not protect the monopolist.

Under existing law regulators can, sometimes, apply antitrust-type criteria in pricing decisions. However, it would be better if state and federal laws governing such commissions were revised to grant state and federal energy regulatory commissions antitrust-like decisional authority (and no more).[20] Such legislation would limit the price actions of public utility commissions to no more than that required for deciding antitrust considerations on price. It also would revise the power to grant exclusive territories into a power to assure a right of competitive entry. If (illegal!) monopolies exist, regulators simply would make it possible for competitors to offer alternative services. If undue price differences exist by the criteria of antitrust law, regulators would require sellers to remove improper cross-subsidies. If unfair (anti-competitive) trade practices are found, regulators would require the parties to eliminate them.

Is Antitrust "Muscular" Enough to Constrain Utilities?

In the absence of a provable antitrust claim, the regulators should do nothing at all. Critics might suggest that antitrust is insufficient to control utilities; that they would still have wide latitude to "gouge"

---

[19]This was in fact the legal background for the rationale of the Supreme Court in upholding regulation of interstate carriers in the 1913 *Pipe Line Cases.*

[20]Both state and federal utility regulatory commissions can to a large degree already fashion remedies that are essentially those that an antitrust court also would require. Regulators cannot award "damages." Only courts of law can do that in the American constitutional system. I do not recommend that utility regulators be given such authority.

113

consumers by various means since traditional antitrust law cannot be used simply as an instrument of price control. But the feared power utilities might have to gouge consumers could only derive from markets that are closed to entry. Antitrust power could therefore remedy the underlying condition that supposedly allows the "gouging" to occur in the first place.[21]

For those who are skeptical that the regulation of energy markets is best viewed as just another exercise in procompetitive legal policy rather than a policy problem that requires a completely different orientation, it would be profitable to review two important cases in regulatory law: the *Pipe Line Cases* of 1913 and the 1943 case of the *Federal Power Commission v. Hope Natural Gas Company*.

The "Pipe Line Cases"

The 1913 Supreme Court opinion called the *Pipe Line Cases* turned on the issue of "tying." Tying exists if the provider of a desired service or product requires that the buyer also purchase another service or product to obtain the desired service. This required purchase is said to be "tied" to the desired purchase. When the desired service or product is also one that gives market power to the seller of that service or product, then the requirement of a tied purchase may be considered an exercise of improper market power. Tying may have the effect of raising the effective price of the desired product while monopolizing the (perhaps otherwise competitive) market for the tied product or service.

State and United States antitrust laws prohibit tying. For example, Section 3 of the Clayton Act stipulates that

> "it shall be unlawful for any person . . . to lease or make a sale or contract for a sale of goods . . . or fix a price charged therefore, or discount or rebate upon, such price, on the condition, agreement or understanding that the lessee or purchaser thereof shall not use nor deal in the goods of a competitor . . . where the effect . . . may be to substantially lessen competition or to tend to create a monopoly."

---

[21]It should be noted, however, that the mere presence of some bottleneck to competition does not necessarily justify regulatory action. The excess profits earned by those positioned to take advantage of bottlenecks invariably attract new entrants who then work to lower prices.

In the *Pipe Line Cases* the power of the federal government to regulate transportation on railroads or pipelines was upheld. The case dealt with transportation by pipelines owned by Standard Oil Company and attached to petroleum fields in which producing properties were owned by both Standard Oil and others. The Court offered the following rationale:

> Availing itself of its monopoly of transportation the Standard Oil Company refused through its subordinates to carry any oil unless the same was sold to it or to them and through them to it on terms more or less dictated by itself. In this way it made itself the master of the fields without the necessity of owning them and carried across half the continent a great subject of international commerce coming from many owners but, by the duress of which the Standard Oil Company was the master, carrying it all as its own.[22]

Although the term was not applied in the decision, the Court's rationale for allowing regulation was the existence of tying used for anti-competitive effect. The monopolized market was the pipeline service. The operator of the pipeline was accused of tying the use of that service to the required sale of oil to be transported. The provision of transport was used to monopolize a market that otherwise would have been competitive.

> The statute practically means no more than that they [the Standard Oil Company] must give up requiring a sale to themselves before carrying the oil that they now receive.[23]

The Supreme Court found in 1913 that the regulator whose actions were in question (the Interstate Commerce Commission, which was acting as a traditional-style regulator, not as an antitrust enforcement agency) could be empowered by Congress to regulate transportation rates on interstate pipelines, partly because the matter regulated was something we today regard as an antitrust or anti-competitive violation.

Nor was this finding likely to have been a surprise to the participants in the case. Similar questions were clearly on the mind of Congress. After granting the relevant regulatory authority to the

[22]234 U.S. at 559.
[23]234 U.S. at 561.

Interstate Commerce Commission in 1906 in the *Hepburn Act*, Congress asked the newly empowered Commission to report to Congress on results of the investigation of transport problems. On January 29, 1907, the Commission reported to the Congress that "More than anything else, the pipeline has contributed to the monopoly of the Standard Oil Company." The Commission then invited parties seeking to transport on the pipeline to file complaints with the Commission upon which it might then act against Standard Oil.[24]

The Hope *Case*

A second frequently cited regulatory case decided by the United States Supreme Court is often referred to as *Hope*.[25] The Hope Natural Gas Company was unhappy with a decision of the Federal Power Commission regarding the total value of plant allowed to be recovered in regulated rates. Companies want more dollars "recovered" in rates while other participants want less. The Commission picked less, Hope wanted more, and when the matter reached the Supreme Court, Hope lost *Hope*. The case is cited by applicant companies in regulatory proceedings often because the Court used language that, if taken out of the entire context of the case, seems to favor strongly the full recovery of claimed costs. The company is entitled to a return:

> sufficient to assure confidence in the financial integrity of the enterprise, so as to enable it to maintain its credit and to attract capital [and to] enable the company to operate successfully, to maintain its financial integrity, to attract capital, and to compensate its investors for the risks assumed.[26]

But if regulated companies like this language of *Hope*, they often forget to cite a case decided only one year later in which the Supreme Court specifically addressed the meaning of those words that seem strongly to favor always granting cost recovery sought by applicants. In that 1944 case, the regulated entity in question was the Market

---

[24]"Railroad Discriminations and Monopolies in Coal and Oil. Letter from the Chairman of the Interstate Commerce Commission Submitting a Report of an Investigation of the Subject of Railroad Discriminations and Monopolies in Oil," Document No. 606 to the House of Representatives, 59th Congress, 2nd Session, p. 14.

[25]*Federal Power Commission, et al., v. Hope Natural Gas Company*, 302 U.S. 591 (1944). One suspects more than happenstance that this nickname is also a pun.

[26]320 U.S. 591 at 603 and 605.

Street Railway Company, which had operated a system of passenger rail service in and near San Francisco. The Railroad Commission of California had denied a request for a certain amount of depreciation to be included in the rates of the company. By the time the appeal reached the U.S. Supreme Court, the Market Street Railway was in bankruptcy. The Market Street Railway Company thought the language of the *Hope* case was particularly compelling for its own situation.

The Supreme Court interpreted its 1943 decision in *Hope* in its 1944 decision in *Market Street Railway*[27] as follows:

> It was noted in the *Hope Natural Gas Case* that regulation does not assure that the regulated business make a profit. . . . All that was held was that a company could not complain if the return which was allowed made it possible for the company to operate successfully. There was no suggestion that less might not be allowed when the amount allowed was all that the company can earn. . . . The due process clause has been applied to prevent governmental destruction of existing economic values. It has not and cannot be applied to insure values or to restore values that have been lost by the operation of economic forces.[28]

And furthermore,

> Even monopolies must sell their services in a market where there is competition for the consumer's dollar and the price of a commodity affects its demand and use.[29]

It is no wonder that lovers of *Hope* avoid *Market Street*! Clearly, these 1913, 1943, and 1944 cases show that the idea that law requires regulation to respect, not replace, competition is not new.

The Need for Legislative Action to Adopt the Antitrust Option

If commissions already can consider competition matters, in at least some form, why must state and federal laws governing such commissions be revised to limit regulatory commissions only to

---

[27]*Market Street Railway Co. v. Railroad Commission of California, et al.,* 324 U.S. 548 (1944).

[28]324 U.S. at 566 to 567.

[29]324 U.S. at 569.

antitrust-type actions? There are several reasons. One is that in many if not most state or federal energy regulatory jurisdictions, antitrust cases would not come before the energy utility commission at all. When competition issues arise at an energy regulatory commission, it almost always is in the context of deciding other matters, such as the depreciation or rates matters cited in other cases mentioned. Further, many commissions now are not required to engage such issues at all. Even when the Supreme Court seems to have dealt with the question, parties can find ways to avoid it—as the discussion of how parties like to cite the *Hope* and avoid the *Market Street* cases illustrates.

The lack of a clear mandate for state regulators to deal with energy issues as questions of antitrust also leads to rather confused case law when regulated companies go to court. For example, two major energy companies recently brought a dispute in a Wyoming federal district court. Colorado Interstate Gas Company alleged that Natural Gas Pipeline Company of America had restricted access to the Natural pipeline in the interstate system.[30] Based on reasoning behind the 1913 *Pipe Line Cases*, one would think that the outcome of the case also would be decided on some criteria related to competition. That was not the result. Rather, the appeals court tossed out a lower court award of damages for competitive harm. The court acted in part on the fact that the challenged tariffs were regulated by the Federal Energy Regulatory Commission.

A similar finding was applied by a federal district court in Illinois, when the state of Illinois among others brought a competition complaint against an interstate pipeline. The decision in that case made clear that the judge may have believed that anti-competitive behavior existed but that, in any event, the rates and actions in question were under the jurisdiction of the Federal Energy Regulatory Commission. Thus, no decision against the company for anti-competitive violations was entered by the court.[31]

Although courts apparently may defer to the Federal Energy Regulatory Commission on matters of competition affecting regulated

[30]*Colorado Interstate Gas Company v. Natural Gas Pipeline Company of America et al.,* 1989 U.S. App., LEXIS 13559.

[31]*State of Illinois et al. v. Panhandle Eastern Pipe Line Corporation.* Decision of Judge Michael M. Minn in docket No. 84-1048, _____.

companies, the Commission does not always apply antitrust considerations to its ratemaking process. The Commission can use such considerations (and at times such as cited elsewhere in this text has done so), but it is not required to do so. In the narrowest construction, the Commission could argue that if it encounters what is believed to be an antitrust violation, it is required to refer the matter to the federal Department of Justice. Most state utility regulators could probably act in a similar manner.

If this confusion were not enough, consider the *Wyoming Tight Sands* case. The dispute arose because federal regulation in the early 1980s had changed the rules of how prices for natural gas production should be set. Certain holders of affected contracts modified them accordingly, but with some question as to the exact date at which they did so. Utilities affected by the altered prices sued the original parties. State regulators eventually joined the fray, but the court determined that the state regulator—the Kansas State Corporation Commission—was not even a properly interested party, despite the fact that the very prices and contracts in question had earlier come before the Kansas Commission!

Had prices never been regulated by utility commissions, none of these (actual or alleged) events, or lawsuits, would have occurred. Instead, under the antitrust option proposed here, the proper regulatory questions from the outset would involve the degree to which competitive markets have been injured.

Conclusion

There is no economic reason to regulate energy by special laws or commissions at all. But the history of the politics of energy suggests that it may be difficult to remove all such regulation and simply leave energy subject to the same laws as the rest of the economy. The best course is certainly to correct this classical historical error and desist from further political treatment of energy. But if state legislatures and the Congress are again unwilling to restrain themselves, there is a second best course that does the least damage: to remove all existing state and federal regulatory commission laws authorizing them to limit entry or regulate price, and instead convert the regulatory commissions into no more than specialized courts for energy antitrust.

7. Energy Policy Reconsidered

Many of the ideas that justify and govern traditional American energy regulation are simply wrong. Energy prices are not doomed to rise forever. Energy supply is not about to dwindle away and disappear forever. Instead, energy exploration, production, and distribution are just technological products. The economics of energy are strongly driven by the economics of discovery, invention, and general progress. Natural monopolies are figments of regulatory imagination. It is time we allowed the ordinary forces of competition to take their proper role in the energy industry.

Drawing Conclusions

This book grew from a private study begun in 1989 related to strategic planning for a utility company. In the course of that study it became clear that, contrary to popular wisdom, natural gas was not in inherently short supply; that its price would fall in the short term and continue to be low in the longer term; and that companies willing to pay the market price for natural gas could likely get as much as they wanted at prices comparable to those of other fuels.

In 1989 that conclusion was considered by many as unbearably radical. Today, only a few years later, it seems almost uncontroversial. Mere mention that energy production is a technologically driven process no longer induces (as it literally did then) hoots of derision. But traditional views are deeply embedded. Recommendations made in this book for the removal or reorientation of traditional regulatory institutions in the United States may seem as radical as was the "technology drives price" argument in those ancient days of 1989. I can ask only for rational review.

The idea that technology is related to energy price, or at least affects energy cost, is today so well known that even cautious trade journals such as the widely read weekly *Oil and Gas Journal* now run frequent articles on the subject.[1] Appropriately to its interests, the petroleum industry sees the issue as largely, but not entirely, one of price. For example, in the last editorial of 1993, entitled "Has crude oil value taken a step down?", the *Oil and Gas Journal*[2] observed, "A crucial question is how much of the current price slump results from transitory factors and how much results from costs permanently lowered by technology." The editorial concluded that "The implications are enormous for producers, consumers, and policy makers [with] . . . non-petroleum agendas."

That conclusion is exactly right. This book does not promote use or avoidance of any particular fuel. If opening energy markets results in increased use of natural gas, as many would likely guess today, proliferation of many local solar-powered communities, expanded construction of newer design nuclear powered plants with reduced total plant plus energy costs, or any other result, so be it.

That is why this book is not only about energy. It is also about a long-standing failure of American political culture. Energy markets have been regulated for more than 100 years. Had the political and academic arguments justifying this regulation been correct, we long ago would have run out of energy. But the critical intellectual processes of American government and of the academy were paralyzed by the disease of collective thinking. This disease, though weakened by political events in the 1980s, remains in place in the United States. It remains capable of enforcing demonstrably false technical ideas as "scientific policy."

---

[1]One recent journal article even contains several graphs of energy price forecast errors that are nearly identical to those used in my private seminars in 1990, and which were at the time literally hooted down by managers of more than one major producer in the natural gas industry. The arguments made are also remarkably similar to my own publications from that period, and especially in Paul Ballonoff and Diana Moss, "Natural Gas Forecasting: One Explanation for Upward Bias," in *Energy Exploration and Exploitation*, vol. 9, no. 3, 1991, pp. 107–21. I created the technology arguments in earlier client studies. Dr. Moss then created the graphs used in that article, subsequently explored issues of price elasticity of energy supply, and prepared similar graphs of erroneous energy supply forecasts—matters also discussed in the recent publication mentioned. See William L. Fisher, "How Technology Has Confounded U.S. Gas Resource Estimators," *Oil and Gas Journal*, October 24, 1994, pp. 100–107.

[2]December 27, 1993 issue, vol. 91, no. 52, p. 19.

The Intellectual Tyranny of the Status Quo: Some Final Questions

If the finite resource model for energy does not match reality, why does the model continue to underpin legislation and regulation? If rising bubble price forecasts are wrong, why are they historically almost the only price forecasts ever published by government or academic institutions? What sustains the ongoing false idea of a crisis that never quite happens?

One true, but overly simple, answer is that the ideas are maintained because it serves the interests of their proponents to maintain them. That is hardly remarkable insofar as it describes private parties—one expects rational private entities to assert their self-interest. But we expect more from academic institutions and government agencies; we instead expect them to seek scientific truth, or at least to project a more dispassionate view of the public interest.

It is easy to understand why some private energy companies might be happy to have a "scientific" foundation for something they might wish for anyway: perpetually higher prices and a market willing to pay them. Distributors may want to raise their own prices, whether to pay their labor unions or their shareholders or their other costs. It is very convenient for utility companies to be able to justify cost increases to regulators with a claim that prices must necessarily go up, sharply, soon, because science says that someone else is responsible: energy is a finite resource and, therefore, those other guys, the producers (not us mere distributors) create a shortage of this finite stuff and force us into these perpetual price increases!

Energy producing companies, which may be maligned by these assertions, also may not be too troubled when utilities tell similar stories to their state regulatory commissions. There are at least three reasons for this. First, the producing companies are not themselves regulated by the state commissions, so the fable has no immediate effect on them. Second, if the state commissions allow price increases based on such arguments, then producers might also conclude that they could in fact raise prices and have them successfully passed through to consumers—thus incidentally proving the argument correct! Finally, producers might be telling the same fable to their bankers. Bankers after all are a cautious lot who want assurances that the funds they lend can be repaid; if it is scientific truth that the price is necessarily going to rise, sharply, soon, then the banker should be more willing to lend larger amounts and on more favorable terms.

This is not to say that all energy companies operate in such a world of fantasy. Quite the opposite: many have been rather honestly analytical in public. And no company will long survive if its business planning is based on a fantasy about price and supply that has not happened in a century, despite what the public relations department, the regulatory affairs department, or the financial affairs officer might prefer to tell their respective publics.

It is fairly easy to understand why government also might like the finite resource model for energy resources. A primary function of government is to repair crises. The finite resource model continually predicts that, no matter what the present condition, energy crises will soon occur that government might help postpone, with just a little more planning now. Thus, the finite resource model as a justification for regulation inserts government permanently into the energy business. The finite resource model is ideally suited for use as a political agenda for an expanding government.

The real issue is why academic or government analysts have not long ago challenged the finite resource model of long-term energy supply and price based on the obvious fact that its predictions are always wrong.[3] In the first place, regulation of energy, especially by the states, did not arise because people conducted analyses that led them to believe resources were finite. Rather, state utility regulation initially arose because local services were vital to the voter's purse and to the creation of bases for local political power. Politicians responded as only politicians can: create a separate committee (that is, a commission) so that someone else will take the political heat. Once a few states proved the idea was legally viable, the federal government also pushed the idea along. Congress created its own early regulators, such as the Interstate Commerce Commission, created by The Act to Regulate Commerce in 1887.

---

[3]An interesting exception, apart from articles by the present author, is a 1992 working paper of an institute at the Massachusetts Institute of Technology. That study concluded that the existence of pervasive consistently wrong forecasts that oil prices will grow at 3 percent annually "represents the result of an exogenous assumption, not the result of analysis." See M. C. Lynch, "The Fog of Commerce: The Failure of Long-Term Oil Market Forecasting," unpublished paper, September 1992.

Congress also encouraged states to form regulatory bodies by placing clauses in territorial rail rate legislation that devolved federal rate authority to new states as soon as they were able to assume that duty.[4] The ability of local politicians to assume such regulatory authority probably encouraged more rapid formation of state governments for the territories, as it provided a source of effective local political power for the newly formed governments. It also provided the base for more secure federal authority in the western expansion. Thus, despite the appearance of separation of powers in our federal framework, there has been a long history of the expansion of federal authority through promotion of regulatory policy at the state level.

But regulatory decisions by commissions could not arise without factual foundations. Making decisions required factual hearings similar to those today that typically precede commission actions. Rationales such as the existence of natural monopoly or energy as a finite resource were required only after commissions were in existence, able (and needing) to write orders that could justify their decisions. Facts followed politics.

The modern federal intellectual control of American business, science, and academic institutions ought to cause serious rethinking of the entire reach of federal "general welfare" powers. Mere claims of the "public good" do not justify federal exercise of a power not enumerated to the federal government. Merely following "due process" before performing an otherwise unconstitutional power to act does not itself confer that power. When all facts are political, then truth has no meaning. When truth has no meaning then law is replaced by raw exercise of power. Surely that result was not the intent of those who wrote our Constitution.

---

[4]See discussion of Supreme Court at pp. 421–23 in its decision in *The Minnesota Rate Cases*, 230 U.S. 352 (1912).

Index

About the Author

Paul Ballonoff holds an A.B. in economics, and M.A. and Ph.D. degrees in social anthropology from the University of California at Los Angeles, and a J.D. from Northwestern California University School of Law (admitted, State Bar of California). Since 1991 he has operated a consulting service providing regulatory and legal analysis, strategic planning, pricing, and related services to the energy industry. He has advised the World Bank, the Inter-American Development Bank, the Philippine Energy Regulatory Board, state regulator boards and commissions, and many domestic and international corporations from all parts of the energy industry. He resides in Alexandria, Virginia.

Cato Institute

Founded in 1977, the Cato Institute is a public policy research foundation dedicated to broadening the parameters of policy debate to allow consideration of more options that are consistent with the traditional American principles of limited government, individual liberty, and peace. To that end, the Institute strives to achieve greater involvement of the intelligent, concerned lay public in questions of policy and the proper role of government.

The Institute is named for *Cato's Letters*, libertarian pamphlets that were widely read in the American Colonies in the early 18th century and played a major role in laying the philosophical foundation for the American Revolution.

Despite the achievement of the nation's Founders, today virtually no aspect of life is free from government encroachment. A pervasive intolerance for individual rights is shown by government's arbitrary intrusions into private economic transactions and its disregard for civil liberties.

To counter that trend, the Cato Institute undertakes an extensive publications program that addresses the complete spectrum of policy issues. Books, monographs, and shorter studies are commissioned to examine the federal budget, Social Security, regulation, military spending, international trade, and myriad other issues. Major policy conferences are held throughout the year, from which papers are published thrice yearly in the *Cato Journal*. The Institute also publishes the quarterly magazine *Regulation*.

In order to maintain its independence, the Cato Institute accepts no government funding. Contributions are received from foundations, corporations, and individuals, and other revenue is generated from the sale of publications. The Institute is a nonprofit, tax-exempt, educational foundation under Section 501(c)3 of the Internal Revenue Code.

CATO INSTITUTE
1000 Massachusetts Ave., N.W.
Washington, D.C. 20001